T0276068

# UNITEXT - La Matematica per il 3+2

Volume 95

**Editor-in-chief**

A. Quarteroni

**Series editors**

L. Ambrosio
P. Biscari
C. Ciliberto
M. Ledoux
W.J. Runggaldier

More information about this series at http://www.springer.com/series/5418

Walter Gander

# Learning MATLAB

A Problem Solving Approach

 Springer

Walter Gander
Departement Informatik
ETH Zürich
Zürich
Switzerland

ISSN 2038-5722          ISSN 2038-5757   (electronic)
UNITEXT - La Matematica per il 3+2
ISBN 978-3-319-25326-8          ISBN 978-3-319-25327-5   (eBook)
DOI 10.1007/978-3-319-25327-5

Library of Congress Control Number: 2015953792

Mathematics Subject Classification (2010): 65-XX

Printed on acid-free paper

Springer International Publishing AG Switzerland is part of Springer Science+Business Media
(www.springer.com)

# Acknowledgment

I would like to thank Dr. Karl Knop for his interest in this project, for proofreading several chapters and for solving problems and exercises.

# Contents

# Introduction

## How to Use This Book

The goal of this book is to teach MATLAB by examples, that is, by showing how to solve problems by designing an algorithm and implementing it in MATLAB. Cleve Moler developed MATLAB originally for teaching linear algebra. MATLAB is the acronym for "MATrix LABoratory." Today MATLAB is a widely used computer language for technical computing. This book is not meant to cover the whole range of MATLAB. Rather it is an introduction to motivate the students to learn this programming language.

The book is based on notes that have been written for a beginner course of 7 weeks with 3 hours of lectures and exercises per week, given at Qian Weichang College at Shanghai University in the fall of 2014. I am indebted to Ying Su for her help during that course and to Prof. Chuan-Qing Gu who invited me to lecture this course.

Some examples were taken from the books [3], [4] and from the freely available online book of the MATLAB creator [8].

Most programs developed in this course can also be run using the public domain software GNU OCTAVE.[1] The OCTAVE language is quite similar to MATLAB so that programs are portable.

Programming environments like MATLAB are very large systems. Getting familiar with the graphical user interface of MATLAB is for a beginner already a challenge. Chapter 1 gives a few hints how to get organized.

It is not possible to get familiar with the whole MATLAB system in one semester, not even in several semesters! My approach is therefore based on *learning by doing*. Given a problem, one has to find a way to solve it using MATLAB. My experience is that the students memorize much better MATLAB commands and programming structures when they *use it themselves*. Therefore it is important to do the exercises before consulting Chap. 10 where all solutions are given.

---

[1]http://www.gnu.org/software/octave/.

Several topics of the book are taken from my area of interest: Scientific Computing. The emphasis is, however, on programming. Showing how to compute some elementary functions using the four basic operations (Chap. 4) is just a nice programming exercise. Computations are mostly done with the usual IEEE arithmetic implemented in MATLAB. Chapter 2 describes the principles of this finite arithmetic. In Chap. 5 we use some of MATLAB's integer arithmetic operations by working with unsigned integers. We do not make use of any of the many toolboxes of MATLAB, and especially we do not use Symbolic Math Toolbox Functions. It is important to tell the students the difference when computing with a computer algebra system versus using standard IEEE arithmetic.

The other book which also teaches MATLAB but focuses more on scientific computing is [9].

Zürich, Switzerland                                          Walter Gander
Summer 2015

# Some Historical Remarks on the Genesis of MATLAB

Linear Algebra, especially matrix algebra, is of the utmost importance for scientific calculations, as the solutions of many problems are constructed from fundamental operations in this field. Nonlinear problems are often solved iteratively in such a way that in each iteration step one has to solve a linear problem. These are essentially matrix operations, solving linear systems of equations and eigenvalue problems. This fact was recognized early on, so already in the 1960s, a program library for linear algebra was being constructed. At that time, scientific computing was performed exclusively in two programming languages, ALGOL 60 and FORTRAN. A series entitled "Handbook for Automatic Computation" was started by the Springer publishing company, so as to, one day, obtain a complete and reliable library of computer programs. The documentation language was defined to be ALGOL, as

> indeed, a correct ALGOL program is the abstractum of a computing process for which the necessary analyses have already been performed.[2]

The first volume of the handbook consists of two parts: In the first part A, H. Rutishauser describes the reference language under the title "Description of ALGOL 60" [10], in the second part B "Translation of ALGOL 60," the three authors Grau, Hill, and Langmaack [6] provide instructions on how to build a compiler. We have to be aware that in those times the computers were delivered with almost no software!

The second volume of the handbook, edited by Wilkinson and Reinsch, appeared in 1971. Under the title of "Linear Algebra" [12], it contains various procedures to solve linear systems of equations and eigenvalue problems. The quality of this software is so good that the algorithms are still used today.

Unfortunately, this series of handbooks was discontinued, as the fast development and distribution of information technology made any further coordination impossible.

---

[2]Heinz Rutishauser in [10].

Because of the language barrier Europe—USA

The code itself has to be in FORTRAN, which is the language for scientific programming in the United States.[3]

the LINPACK [2] project was executed in 1970s at Argonne National Laboratory. LINPACK contains programs for the solution of fully occupied systems of linear equations. They were based on the procedures of the handbook, but had been renewed and systematically programmed in FORTRAN. This can be seen in the unified conventions for naming, portability, and machine independence (e.g., termination criteria), use of elementary operations by calling the BLAS (Basic linear Algebra Subprograms). The LINPACK Users Guide appeared in 1979. LINPACK is also the name of a benchmark for measuring performance of a computer in floating point operations. This benchmark used to consist of two parts: On the one hand, a given FORTRAN program had to be compiled and executed to solve a fully occupied $100 \times 100$ system of linear equations; on the other hand, a $1000 \times 1000$ system of linear equations had to be solved as fast as possible (using any adjusted program). This benchmark,[4] in a modified form, is now used every 6 months to determine the 500 most powerful computers in the world, so as to list them in the top 500 list, see http://www.top500.org.

Also the eigenvalue procedures from [12] were translated into FORTRAN and are available under the name EISPACK [5], [11]. EISPACK and LINPACK were replaced a number of years ago by LAPACK [1]. The LINPACK, EISPACK, and LAPACK procedures (and many more) can be obtained electronically from the online software library NETLIB, see http://www.netlib.org.

In the late 1970s, Cleve Moler developed the interactive program MATLAB (MATrix LABoratory), initially only to provide a simple calculation tool for lectures and exercises. The basis for this program, were programs from LINPACK and EISPACK. As efficiency considerations were not of great importance, only eight procedures from LINPACK and five from EISPACK for calculations with full matrices were included. MATLAB was not only established as a useful teaching aid, but also applied in contrary to the initial intention, in research and industry. The initially public domain MATLAB [7], written in FORTRAN was completely overworked, extended and made it into an efficient engineering tool by the company MathWorks.[5] It is now written in C. This philosophy in the development of MATLAB has led to a continuous writing of new function packages (so-called Toolboxes) for various fields of application. The user community has a discussion platform at MATLAB-Central http://www.mathworks.com/matlabcentral/?s_tid=gn_mlc_logo with a lot of useful information.

---

[3]Citation from the preface of the LINPACK users guide.

[4]https://en.wikipedia.org/wiki/LINPACK_benchmarks.

[5]http://www.mathworks.com.

# Chapter 1
# Starting and Using MATLAB

## 1.1 Organize Your Desktop

Get rid of not necessary open windows on your computer. We need only two windows: one window for executing MATLAB programs and a second one for writing programs. Make these windows as high and large as possible so that you can have a good overview of your programs and of the results. After starting MATLAB it is a good idea to write the command `format compact`. This command eliminates blank lines and thus concentrates the output.

Notice that MATLAB programs are written as plain ASCII texts. You can use any editor to write them. If you use a Linux operating system then *Emacs* or *vi* are very good choices.

## 1.2 MATLAB Scripts and Functions

You will write in MATLAB your own programs and your own functions. It is important to distinguish between a program (or MATLAB script) and a function. MATLAB scripts are "main programs" and functions which you write can be used in them. The functions must be stored in the same directory where the script is which calls them. A MATLAB-script is stored as a M-file (a file with the suffix `.m`) and is executed in the command window by typing its name without the suffix.

Notice that you may want to store your programs in a special directory not necessarily the default directory where MATLAB is called. For this you can use the command `cd` which is the abbreviation for "change directory". For instance you may want to write your programs in a directory called `c:\LinearAlgebraProblems`. Then after calling MATLAB you can change directory in the command window with

© Springer International Publishing Switzerland 2015
W. Gander, *Learning MATLAB*, UNITEXT - La Matematica per il 3+2 95,
DOI 10.1007/978-3-319-25327-5_1

1

```
>> cd c:\LinearAlgebraProblems
```

if you want to see what files are in that directory then use

```
>> ls
```

it will give you a list of all the files.

## 1.2.1   MATLAB *Script*

Assume that you are given a linear equation

$$ax + b = 0$$

and you would like to write a program to solve such equations. You could write the following text and save it as an M-file:

```
% solves the linear equation a*x+b=0 for x.
 a=input('a=?')
 b=input('b=?')
 if a~=0,
    x=-b/a
 elseif b==0,
    disp('any x is solution')
 else
    disp('no solution')
 end
```

Now save this program under the name `LinearScript.m` and call this program in the MATLAB command-window to solve the equation $3x + 5 = 0$

```
>> LinearScript
a=?3
a =
      3
b=?5
b =
      5
x =
   -1.6667
```

## 1.2.2   MATLAB *Function*

A function can be called by different MATLAB scripts and also by other functions. A function has input parameters and delivers results as output parameters. So for our linear equation the input parameters are the two coefficients $a$ and $b$ and the output parameter, the result, is the solution $x$. The function becomes

```
function x=SolveLinear(a,b)
% SolveLinear solves the linear equation a*x+b=0
if a~=0,
   x=-b/a;
elseif b==0,
   error('any x is solution')
else
   error('no solution')
end
```

Save this function under the name `SolveLinear.m`. *Note that the file name must be the same as the function name*. Now in order to apply the function we can write in the MATLAB window

```
>> SolveLinear(3,5)
ans =
   -1.6667
```

Or we could type another MATLAB script interactively in the MATLAB window

```
>> b=1;
>> for a=-3:2
   a
   x=SolveLinear(a,b)
end
a =
   -3
x =
   0.3333
a =
   -2
x =
   0.5000
a =
   -1
x =
   1
a =
   0
??? Error using ==> SolveLinear at 8
no solution
```

the program stops execution because $a = 0$.

It is better, however, is to write and save the script, say with the name `mainLinear.m`. Then execute it in the MATLAB-command window:

```
>> mainLinear
a =
   -3
x =
   0.3333
a =
   -2
x =
   0.5000
```

```
a =
     -1
x =
     1
a =
     0
Error using SolveLinear (line 8)
no solution
Error in mainLinear (line 4)
  x=SolveLinear(a,b)
```

we get essentially the same output but it is more convenient to correct errors or
change the program.

## 1.3   The Windows Environment

To run MATLAB on a PC, double-click on the MATLAB icon. This will present the
following screen

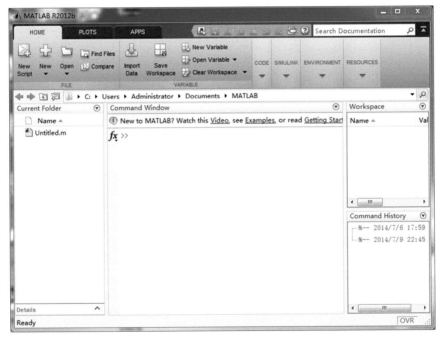

The working directory will depend on where MATLAB has been installed. To
check in which directory you are, use the command pwd which is the abbreviation
for "print working directory". You then maybe get

```
>> pwd
ans =
  C:\Users\Administrator\Documents\MATLAB
```

or some similar information. To work in the directory

```
C:Users\myMatlabPrograms
```

use the command `cd` which is the abbreviation for "change directory":

```
>> cd C:Users\myMatlabPrograms
>> pwd
ans =
  C:Users\myMatlabPrograms
```

It is very convenient to work with a script window where you write your programs. Write and store the program under some file name for instance `LinearScript.m`. Then call it in the MATLAB window

```
>> LinearScript
```

to execute it and get the results.

To quit MATLAB at any time, type `quit` or `exit` at the MATLAB prompt. If you feel you need more assistance, you can:

- Access the Help Desk by typing `doc` at the MATLAB prompt.
- Type `help <subject>` at the MATLAB prompt, for instance if you like to know more about functions you would type `help function`.
- Pull down the **Help** menu on a PC.

## 1.4   The Linux Environment

In a shell you call `matlab` and MATLAB will present the following screen

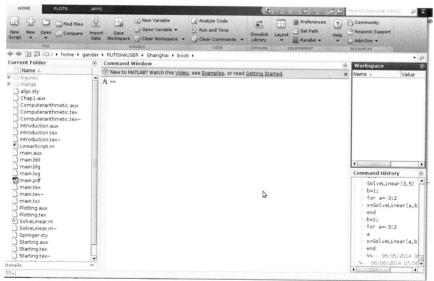

If you don't want to work with the GUI then call `matlab -n` to obtain MATLAB in the same shell without desktop:

```
gander@pnb-502:~$ matlab -n
Starting matlab as: /home/system/opt/matlab/default/bin/matlab  -nodesktop

                    < M A T L A B (R) >
           Copyright 1984-2013 The MathWorks, Inc.
            R2013a (8.1.0.604) 64-bit (glnxa64)
                    February 15, 2013

     ---------------------------------------------------

To get started, type one of these: helpwin, helpdesk, or demo.
For product information, visit www.mathworks.com.

>>
```

## 1.5   Using GNU Octave

MATLAB without GUI looks very similar to GNU OCTAVE. We start it here and call `SolveLinear(3,5)`:

```
gander@pnb-502:~$ octave
GNU Octave, version 3.2.4
Copyright (C) 2009 John W. Eaton and others.
This is free software; see the source code for copying conditions.
There is ABSOLUTELY NO WARRANTY; not even for MERCHANTABILITY or
FITNESS FOR A PARTICULAR PURPOSE.  For details, type 'warranty'.

Octave was configured for "x86_64-pc-linux-gnu".

Additional information about Octave is available at http://www.octave.org.

Please contribute if you find this software useful.
For more information, visit http://www.octave.org/help-wanted.html

Report bugs to <bug@octave.org> (but first, please read
http://www.octave.org/bugs.html to learn how to write a helpful report).

For information about changes from previous versions, type 'news'.

octave:1> SolveLinear(3,5)
ans = -1.6667
octave:2>
```

Newer versions of GNU Octave have also a GUI very similar to MATLAB.

## 1.6   Documenting Results

An easy way to store results generated by computations is to use the command `diary`. Assume you file `LinearScript.m` contains the following lines

```
% solves the linear equation a*x+b=0 for x.
  a=input('a=?')
  b=input('b=?')
  if a~=0,
     x=-b/a
  elseif b==0,
     disp('any x is solution')
  else
     disp('no solution')
  end
```

If you use the `diary` command, you can store the results on a file. Assume you want to store the results on the file `MyResuts.txt`. The you should write

```
>> diary MyResults.txt
>> LinearScript
a=?3
a =
     3
b=?5
b =
     5
x =
   -1.6667
>> diary off
```

The file `MyResults.txt` will be generated in the same directory and contain the session above:

```
LinearScript
a=?3
a =
     3
b=?5
b =
     5
x =
   -1.6667
diary off
```

## 1.7 MATLAB-Elements Used in This Chapter

The description of the MATLAB-elements is taken from the documentation center of the MathWorks Webpage http://www.mathworks.com/.

**format compact:** Suppresses excess line feeds to show more output in a single screen.

| | |
|---|---|
| **M-file:** | executable MATLAB file containing MATLAB-commands. It has to be stored in the same directory where MATLAB is called. You can use the MATLAB-editor or any other text editor to create your .m-files. The files must be stored as plain ASCII text-files with .m-extension. |
| **pwd:** | Identify current folder<br>pwd displays the Matlab current folder. |
| **cd:** | Change current folder<br>cd (newFolder) changes the current folder to newFolder. |
| **ls:** | List folder contents<br>ls lists the contents of the current folder. |
| **quit:** | Terminate MATLAB program |
| **exit:** | Terminate MATLAB program (same as quit) |
| **doc:** | Reference page in Help browser<br>doc name displays documentation for the functionality specified by name, such as a function, class, or block. |
| **help:** | Help for functions in Command Window<br>help name displays the help text for the functionality specified by name, such as a function, method, class, or toolbox. |
| **diary:** | Save Command Window text to file<br>diary('filename') writes a copy of all subsequent keyboard input and the resulting output (except it does not include graphics) to the named file, where filename is the full pathname or filename is in the current MATLAB folder.<br><br>diary off suspends the diary. |
| **if-statement:** | See the script LinearScript.m. It has the form |

```
if expression
   statements
elseif expression
   statements
else
   statements
end
```

| | |
|---|---|
| **%:** | The percent sign is used to comment out a line. Comments are also useful for program development and testing—comment out any code that does not need to run. To comment out multiple lines of code, you can use the block comment operators, %{ and %}. The %{ and %} operators must appear alone on the lines that comment out the block. Do not include any other text on these lines. |

**for:**                  Execute statements specified number of times

```
for index = values
    program statements
        :
end
```

values has one of the following forms:

initval:endval          increments the index variable from initval to endval by
                        1, and repeats execution of program statements until index is
                        greater than endval.

initval:step:endval     increments index by the value step on each iteration, or decre-
                        ments when step is negative.

valArray                creates a column vector index from subsequent columns of
                        array valArray on each iteration. For example, on the first
                        iteration, index = valArray(:,1). The loop executes for
                        a maximum of n times, where n is the number of columns of
                        valArray, given by numel(valArray, 1, :). The input
                        valArray can be of any MATLAB data type, including a string,
                        cell array, or struct.

**input:**                Used to enter input from the keyboard.

                        result = input(prompt) displays the prompt string on
                        the screen, waits for input from the keyboard, evaluates any
                        expressions in the input, and returns the result.

**disp:**                 Used to display things on the screen.

                        disp(X) displays the contents of X without printing the vari-
                        able name.

**function:**             See our function SolveLinear.

                        function [y1,...,yN] = myfun(x1,...,xM)
                        declares a function named myfun that accepts inputs
                        x1,...,xM and returns outputs y1,...,yN.

**error:**                Display message and abort function.

## 1.8  Problems and Exercises

1. Start MATLAB with the GUI and watch the introductory video and study the
   tutorial.
2. If you own a computer or laptop without MATLAB then download and install the
   open source software GNU OCTAVE on it.

# Chapter 2
# How a Computer Calculates

This section is an excerpt of the material presented in Chap. 2 of [3]. It is important to realize that a computer cannot perform numerical computations exactly like one would expect them to be done in mathematics. Understanding its limitations is essential for developing good programs.

## 2.1 Finite Arithmetic

A computer is a finite automaton. This means that a computer can only store a *finite set of numbers* and perform only a *finite number of operations*. In mathematics, we are familiar calculating with real numbers $\mathbb{R}$ covering the continuous interval $(-\infty, \infty)$, but on the computer, we must contend with a discrete, finite set of *machine numbers* $\mathbb{M} = \{-\tilde{a}_{min}, \ldots, \tilde{a}_{max}\}$. Hence each real number $a$ has to be mapped onto a machine number $\tilde{a}$ to be used on a computer.

In addition the finite set of machine numbers $\mathbb{M}$ contains only real numbers with a limited fix number of digits. If this fixed number of digits is 8 then all numbers with the same leading 8 digits will be mapped to the same machine number. The machine numbers are represented as *floating point* numbers (here as example in base 10) that is

$$\tilde{a} = \pm m \times 10^{\pm e}$$

with $m = D.D \cdots D$ the *mantissa*, $e = D \cdots D$ the *exponent* and $D$ is a digit $D \in \{0, 1, \ldots, 9\}$. A nonzero machine number $\tilde{a} \neq 0$ is (to avoid ambiguity) *normalized* which means, that the digit before the decimal point in the mantissa is nonzero. The machine numbers are not spread regularly. They are dense near zero and sparse at the end of the *computation range* $[-\tilde{a}_{min}, \tilde{a}_{max}]$.

If a calculation leads to a result outside the computation range, then we speak of *overflow*. There exists a smallest positive normalized machine number $m$. If we compute $b = m/2$ then we would expect the result to be 0 because there is no normalized machine number between 0 and $m$. Therefore the interval $(-m, m)$ is

W. Gander, *Learning MATLAB*, UNITEXT - La Matematica per il 3+2 95,
DOI 10.1007/978-3-319-25327-5_2

called the *underflow range* (but notice that IEEE arithmetic allow computations with denormalized numbers).

The *machine precision* is traditionally defined to be the smallest machine number *eps* > 0 such that on the computer

$$1 + eps > 1$$

holds. Newer definition just define *eps* as the *spacing of the machine numbers in the interval* [1, 2].

## 2.2  Rounding Errors

Let $\tilde{a}$ and $\tilde{b}$ be two machine numbers; then $c = \tilde{a} \times \tilde{b}$ will in general not be a machine number anymore, since the product of two numbers contains twice as many digits. The computed result will therefore be rounded to a machine number $\tilde{c}$ which is closest to $c$.

As an example, consider the 8-digit decimal machine numbers

$$\tilde{a} = 1.2345678 \quad \text{and} \quad \tilde{b} = 1.1111111,$$

whose product is

$$c = 1.37174198628258 \quad \text{and} \quad \tilde{c} = 1.3717420.$$

The *absolute rounding error* is the difference $r_a = \tilde{c} - c = 1.371742e{-}8$, and

$$r = \frac{r_a}{c} = 1e{-}8$$

is called the *relative rounding error*.

On today's computers, basic arithmetic operations obey the *standard model of arithmetic*: for $a, b \in \mathbb{M}$, we have

$$a \tilde{\oplus} b = (a \oplus b)(1 + r), \tag{2.1}$$

where $r$ is the relative rounding error with $|r| < eps$, the machine precision. We denote with $\oplus \in \{+, -, \times, /\}$ the exact basic operation and with $\tilde{\oplus}$ the equivalent computer operation.

Another interpretation of the standard model of arithmetic is due to Wilkinson. In what follows, we will no longer use the multiplication symbol $\times$ for the exact operation; it is common practice in algebra to denote multiplication without any symbol: $ab \iff a \times b$. Consider the operations

Addition:         $a\tilde{+}b = (a+b)(1+r) = (a+ar) + (b+br) = \tilde{a} + \tilde{b}$
Subtraction:      $a\tilde{-}b = (a-b)(1+r) = (a+ar) - (b+br) = \tilde{a} - \tilde{b}$
Multiplication:   $a\tilde{\times}b = ab(1+r) = a(b+br) = a\tilde{b}$
Division:         $a\tilde{/}b = (a/b)(1+r) = (a+ar)/b = \tilde{a}/b$

In each of the above, the operation satisfies

**Wilkinson's Principle**

The result of a numerical computation on the computer is the exact result with slightly perturbed initial data.

For example, the numerical result of the multiplication $a\tilde{\times}b$ is the exact result $a\tilde{b}$ with a slightly perturbed operand $\tilde{b} = b + br$.

**Cancellation**

A special rounding error is called *cancellation*. If we subtract two almost equal numbers, leading digits are canceled. Consider the following two numbers with 5 decimal digits:

$$\begin{array}{r} 1.2345e0 \\ -1.2344e0 \\ \hline 0.0001e0 = 1.0000e-4 \end{array}$$

If the two numbers were exact, the result delivered by the computer would also be exact. But if the first two numbers had been obtained by previous calculations and were affected by rounding errors, then the result would at best be $1.XXXXe-4$, where the digits denoted by $X$ are unknown.


## 2.3 IEEE-Arithmetic

Since 1985 we have for computer hardware the *ANSI/IEEE Standard 754 for Floating Point Numbers*. It has been adopted by almost all computer manufacturers. The base is $B = 2$. Expressed as decimal numbers this standard allows to represent numbers with about 16 decimal digits and an exponent of 3-digits. More precisely the computation range is the interval

$$\mathbb{M} = [-\tilde{a}_{min}, \tilde{a}_{max}] = [-1.797693134862316e+308, 1.797693134862316e+308]$$

The standard defines *single* and *double* precision floating point numbers.

MATLAB has also adopted IEEE-Arithmetic and computes by default with *double precision*. We shall not discuss the single precision which uses 32 bits.

The IEEE *double precision* floating point standard representation uses a 64-bit word with bits numbered from 0 to 63 from left to right. The first bit $S$ is the sign bit, the next eleven bits $E$ are the exponent bits for $e$ and the final 52 bits $F$ represent the mantissa $m$:

$$\overbrace{\phantom{EEEEEEEEEEE}}^{e} \quad \overbrace{\phantom{FFFFF\cdots FFFFF}}^{m}$$
$$S\ EEEEEEEEEEE\ FFFFF\cdots FFFFF$$
$$0\ 1 \qquad\qquad\quad 11\ 12 \qquad\qquad\qquad 63$$

The value $\tilde{a}$ represented by the 64-bit word is defined as follows:

**normal numbers:**  If $0 < e < 2047$, then $\tilde{a} = (-1)^S \times 2^{e-1023} \times 1.m$ where $1.m$ is the binary number created by prefixing $m$ with an implicit leading 1 and a binary point.

**subnormal numbers:**  If $e = 0$ and $m \neq 0$, then $\tilde{a} = (-1)^S \times 2^{-1022} \times 0.m$ , which are *denormalized* numbers.
If $e = 0$ and $m = 0$ and $S = 1$, then $\tilde{a} = -0$
If $e = 0$ and $m = 0$ and $S = 0$, then $\tilde{a} = 0$

**exceptions:**  If $e = 2047$ and $m \neq 0$, then $\tilde{a} = $ NaN (*Not a number*)
If $e = 2047$ and $m = 0$ and $S = 1$, then $\tilde{a} = -$Inf
If $e = 2047$ and $m = 0$ and $S = 0$, then $\tilde{a} = $ Inf

Using the *hexadecimal format* in MATLAB we can see the internal representation of a floating point number. We obtain for example

```
>> format hex
>> 2
ans =    4000000000000000
```

If we expand each hexadecimal digit to 4 binary digits we obtain the bit pattern for the floating point number 2:

```
 0100 0000 0000 0000 0000 0000 ....  0000 0000 0000
```

We skipped with . . . . seven groups of four zero binary digits. The interpretation is: $+1 \times 2^{1024-1023} \times 1.0 = 2$.

```
>> 6.5
ans =    401a000000000000
```

This means

```
 0100 0000 0001 1010 0000 0000 ....  0000 0000 0000
```

Again we skipped with . . . . seven groups of four zeros. The resulting number is $+1 \times 2^{1025-1023} \times (1 + \frac{1}{2} + \frac{1}{8}) = 6.5$.

From now on, we will stick to the IEEE Standard as used in MATLAB. In other, more low-level programming languages, the behavior of the IEEE arithmetic can be adapted, e.g. the exception handling can be explicitly specified.

- The *machine precision* is $eps = 2^{-52}$.
- The largest machine number $\tilde{a}_{max}$ is denoted by the constant `realmax`. Note that

```
>> realmax
   ans = 1.7977e+308
```

- The *computation range* is the interval $[-\texttt{realmax}, \texttt{realmax}]$. If an operation produces a result outside this interval, then it is said to *overflow*. Before the IEEE Standard, computation would halt with an error message in such a case. Now the result of an overflow operation is assigned the number $\pm\texttt{Inf}$.
- The smallest positive normalized number is $\texttt{realmin} = 2^{-1022}$.
- IEEE allows computations with *denormalized numbers*. The positive denormalized numbers are in the interval $[\texttt{realmin} * \texttt{eps}, \texttt{realmin}]$. If an operation produces a strictly positive number that is smaller than $\texttt{realmin} * \texttt{eps}$, then this result is said to be in the *underflow range*. Since such a result cannot be represented, zero is assigned instead.
- When computing with denormalized numbers, we may suffer a loss of precision. Consider the following MATLAB program:

```
>> format long
>> res=pi*realmin/123456789101112

res =   5.681754927174335e-322

>> res2=res*123456789101112/realmin

   res2 = 3.15248510554597

>> pi    = 3.14159265358979
```

The first result `res` is a denormalized number, and thus can no longer be represented with full accuracy. So when we reverse the operations and compute `res2`, we obtain a result which only contains 2 correct decimal digits. We therefore recommend avoiding the use of denormalized numbers whenever possible.

## 2.4  MATLAB-Elements Used in This Chapter

**eps:**    the machine precision eps returns the distance from 1.0 to the next largest double-precision number, that is eps $= 2^{-52}$.

**realmin:**    Smallest positive normalized floating-point number. n = realmin returns the smallest positive normalized floating-point number in IEEE double precision.

**realmax:**    Largest positive floating-point number.  n = realmax returns the largest finite floating-point number in IEEE double precision.

**format:**        format sets the display of floating-point numeric values to the default
display format, which is the short fixed decimal format. This format
displays 5-digit scaled, fixed-point values.

**format hex:**    Hexadecimal representation of a binary double-precision number.

**NaN:**           Not-a-Number
NaN returns the IEEE arithmetic representation for Not-a-Number
(NaN). These values result from operations which have undefined
numerical results.

**Inf:**           Infinity
Inf returns the IEEE arithmetic representation for positive infinity.
Infinity values result from operations like division by zero and overflow, which lead to results too large to represent as conventional
floating-point values.

## 2.5  Problems

1. Consider the following finite decimal arithmetic: 2 digits for the mantissa and
   one digit for the exponent. So the machine numbers have the form $\pm Z.ZE \pm Z$
   where $Z \in \{0, 1, \ldots, 9\}$

   (a) How many normalized machine numbers are available?
   (b) Which is the overflow- and the underflow range?
   (c) What is the machine precision?
   (d) What is the smallest and the largest distance of two consecutive machine
       numbers?

2. Solving a quadratic equation: Write a MATLAB function

   ```
   function [x1,x2]=QuadraticEq(p,q)
   ```

   which computes the real solutions of an equation

   $$x^2 + px + q = 0.$$

   If the solutions turn out to be complex then write an error message. Test your
   program with the following examples:

   - $(x - 2)(x + 3) = x^2 + x - 6 = 0$   thus $p = 1$ and $q = -6$.
   - $(x - 10^9)(x + 2 \cdot 10^{-9}) = x^2 + (2 \cdot 10^{-9} - 10^9)x + 2$
     thus $p = 2e-9 - 1e9$ and $q = -1e9$.
   - $(x + 10^{200})(x - 1) = x^2 + (10^{200} - 1)x - 10^{200}$
     thus $p = 1e200 - 1$ and $q = -1e200$.

   Comment your results.

# Chapter 3
# Plotting Functions and Curves

In this section we learn how to define functions in MATLAB and how functions and curves can be plotted.

## 3.1 Plotting a Function

Assume we would like to plot the function $f(x) = 1 + \sin x$ for $x \in [0, 2\pi]$. First we have to program the function. One way to do this is to write a file with the name f.m and store it in the same directory where MATLAB was called. The file f.m looks as follow

```
function y=f(x)
y=1+sin(x);
```

In the MATLAB-window we can evaluate this function

```
>> y=f(0)
y =
     1
>> y=f(pi/4)
y =
    1.7071
```

Notice that pi is a predefined constant in MATLAB.

```
>> pi
ans =
    3.1416
```

also the imaginary unit i or 1i is predefined:

© Springer International Publishing Switzerland 2015
W. Gander, *Learning MATLAB*, UNITEXT - La Matematica per il 3+2 95,
DOI 10.1007/978-3-319-25327-5_3

```
>> i
ans =
   0.0000 + 1.0000i
>> 1i
ans =
   0.0000 + 1.0000i
>> 1i^2
ans =
   -1
```

It is better to use `1i` in order to have the variable `i` free for other purposes.

Since machine number are a finite set we cannot really plot a continuous function. We can only sample the function for say *n* equidistant values of *x* and connect adjacent values by a straight line. Assume we compute $n = 7$ values of our function. In MATLAB we store the values in a vector (that is a one dimensional array). We define the spacing between two function values by *h*:

```
>> h=2*pi/7
h =
    0.8976
>> x=0:h:2*pi
x =
  Columns 1 through 7
        0      0.8976    1.7952    2.6928    3.5904    4.4880    5.3856
  Column 8
    6.2832
```

Notice the construction of this vector x. It has the form

```
x=startvalue : stepsize : endvalue
```

For each value of the vector x we want to evaluate our function. This can be done in MATLAB very compactly by just calling f with the vector x as argument

```
>> y=f(x)
y =
  Columns 1 through 7
    1.0000    1.7818    1.9749    1.4339    0.5661    0.0251    0.2182
  Column 8
    1.0000
```

Now we have computed two vectors x and y and we can connect the function values by the plot command by straight line segments:

```
plot(x,y)
```

we get a new graphic window with the function plot (see first plot in Fig. 3.1). It does not look very good. The spacing between the function values is too large! We can easily improve this by using more function values:

```
x=0:0.01:2*pi;
plot(x,f(x))
```

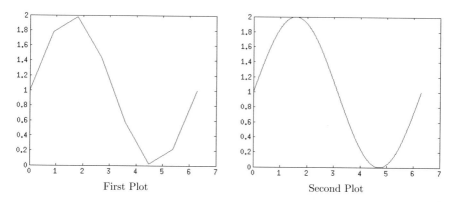

First Plot                                          Second Plot

**Fig. 3.1** Plotting a function with different spacing

Now the plot looks much better (see second plot in Fig. 3.1).

Note that there is a MATLAB-function called `linspace`:

```
>> x=linspace(0,2*pi)
```

it returns in x 100 equidistant values in the interval $[0, 2\pi]$. If you need 500 values you can write `x=linspace(0,2*pi,500)`.

Standard functions in MATLAB can be called with vectors as arguments as we just saw for $f(x) = 1 + \sin(x)$. For a vector $\boldsymbol{x} = (x_1, \ldots, x_n)$ the function $\sin(\boldsymbol{x})$ computes

$$\sin(\boldsymbol{x}) = [\sin x_1, \sin x_2, \ldots, \sin x_n].$$

By adding $1 + \sin(\boldsymbol{x})$ the term 1 is expanded to

```
ones(size(x))+sin(x)
```

so that the result is

$$1 + \sin(\boldsymbol{x}) = [1 + \sin x_1, 1 + \sin x_2, \ldots, 1 + \sin x_n].$$

When we wish to construct functions that allow vectors as arguments we need to use the *dot operations*. These are element by element operations. Consider for instance the function.

$$g(x) = \frac{\sin(x)}{e^x}.$$

Programming this function as

```
function y=g(x)
y=sin(x)/exp(x);
```

would not allow to call $g(x)$ with a vector as argument. However, if we program $g$ as

```
function y=g(x)
y=sin(x)./exp(x);
```

then the elements of the two vectors are divided element-wise and the result is what we want.

## 3.2   Plotting Curves

If we wish to plot a circle with center $C = (c_1, c_2)$ and radius $r$ we need to know the equation of the circle. In Cartesian coordinates the equations reads

$$(x - c_1)^2 + (y - c_2)^2 = r^2.$$

Now to plot the circle in this representation we would have to solve the equation for $y$ and we would obtain two functions $y(x)$. Then we need to compute say 30 $x$-values in the interval $(c_1 - r, c_1 + r)$ and plot the two half-arcs for the two values of $y$.

There is a much simpler solution. We represent the same circle using a parameter $t$ by

$$x(t) = r \cos t$$
$$y(t) = r \sin t, \quad t = 0, \dots, 2\pi.$$

Then we write

```
clear,clf
r=2
C=[0.5,1]
t=linspace(0,2*pi,60)
axis([-3,4,-2,4])
axis equal
hold
plot(C(1)+r*cos(t), C(2)+r*sin(t))
plot(C(1),C(2),'x')
```

to get the plot in Fig. 3.2.

## 3.3   Plotting 3-d Curves

Consider to plot the space curve

$$x(t) = 3 \cos t$$
$$y(t) = 5 \sin t$$
$$z(t) = 2t.$$

**Fig. 3.2** Plotting a *circle* using parametric representation

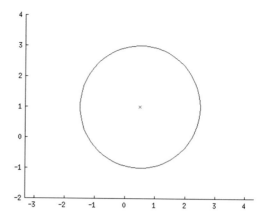

**Fig. 3.3** Plotting a space curve

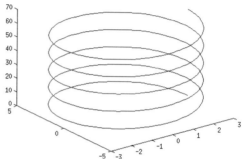

Using the function `plot3` and

```
>> t = 0:pi/50:10*pi;
>> plot3(3*cos(t), 5*sin(t),2*t)
```

we obtain Fig. 3.3.

## 3.4 Surface and Mesh Plots

MATLAB offers many functions for visualizing. See the description on http://www.mathworks.com/help/matlab/surface-and-mesh-plots-1.html. To visualize the surface of a two dimensional function one has to define first a grid in the $xy$-plane on which the function will be evaluated. The grid is defined with the function `meshgrid`. For instance with

```
>>  [x,y] = meshgrid(-1:1:1,0:1:3)
x =
     -1      0      1
     -1      0      1
     -1      0      1
     -1      0      1
```

```
y =
        0        0        0
        1        1        1
        2        2        2
        3        3        3
```

we get a grid of 12 points with the coordinates

$$
\begin{matrix}
(-1,0) & (0,0) & (1,0) \\
(-1,1) & (0,1) & (1,1) \\
(-1,2) & (0,2) & (1,2) \\
(-1,3) & (0,3) & (1,3)
\end{matrix} .
$$

To evaluate the function $f(x, y) = x^2 + y^2$ on that grid we compute

```
>> F=x.^2+y.^2
F =
        1        0        1
        2        1        2
        5        4        5
       10        9       10
```

To plot the surface of the function, MATLAB offers two possibilities. The first one connects the function values in both directions by a mesh:

```
>>   mesh(x,y,F)
```

see the left picture in Fig. 3.4. The second possibility is to connect the Function values by surfaces:

```
>>   surf(x,y,F)
```

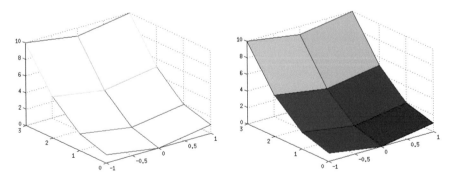

**Fig. 3.4**  mesh and surf function

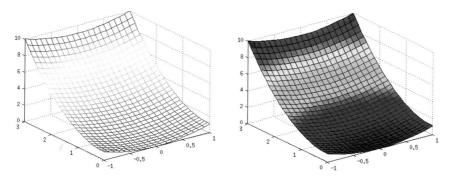

**Fig. 3.5** Finer grid for `mesh` and `surf`

We obtain the right plot in Fig. 3.4. Both figures look nicer if we use a finer grid:

```
[x,y] = meshgrid(-1:0.1:1,0:0.1:3)
F=x.^2+y.^2
mesh(x,y,F)
surf(x,y,F)
```

We get the plots in Fig. 3.5.

## 3.5 Contour Plots

A two dimensional function can also be represented by level lines as a contour plot. Consider the function

$$f(x, y) = \cos(y \cos(x)).$$

To see the level lines of the function in the domain $-\pi \le x \le \pi$ and $0 \le y \le 2\pi$ we program

```
x=linspace(-pi,pi);
y=linspace(0,2*pi);
[X,Y]=meshgrid(x,y);
Z=cos(Y.*cos(X));
figure(1)
contour(X,Y,Z)
figure(2)
mesh(Z)
```

We obtain Fig. 3.6.

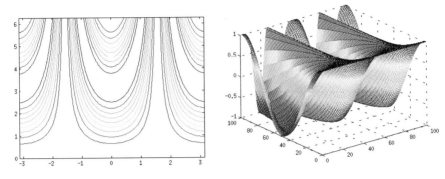

**Fig. 3.6**  Contour and mesh plot

## 3.6  MATLAB-Elements Used in This Chapter

**pi:**               Predefined constant $\pi = 3.141592653589793$.

**1i, i:**            the imaginary unit. It is recommended to use 1i since i may be
                      overwritten and used as variable.

**clf:**              Clear current figure window.

**clear:**            clear removes all variables from the current workspace, releasing
                      them from system memory.

**axis:**             Axis scaling and appearance.

                      axis([xmin xmax ymin ymax]) sets the limits for the x- and y-axis
                      of the current axes. axis equal sets the aspect ratio so that the data
                      units are the same in every direction. This is important to let a
                      circle not appear as an ellipse!

**plot:**             We have used here the simplest form

                      plot(X,Y) creates a 2-D line plot of the data in Y versus the
                      corresponding values in X.
                      There are many useful other MATLAB-function available, see
                      http://www.mathworks.com/help/matlab/line-plots.html.
                      A whole gallery of plot possibilities is discussed on  http://www.
                      mathworks.com/discovery/gallery.html?s_tid=abdoc_plot.

**colon (:) :**       Create vectors, array subscripting, and for-loop iterators

                      The colon is one of the most useful operators in MATLAB. It can
                      create vectors, subscript arrays, and specify for iterations.
                      A linearly spaced vector can be generated by

```
x=startvalue : stepsize : endvalue
```

**linspace:**     This function generates linearly spaced vectors.

y = linspace(a,b) generates a row vector y of 100 points linearly spaced between and including a and b.
y = linspace(a,b,n) generates a row vector y of n points linearly spaced between and including a and b.

**ones:**     Create array of all ones. Example A=ones(3,2) creates a $3 \times 2$ matrix with all elements 1.

**size:**     Array dimensions

d = size(X) returns the sizes of each dimension of array X in a vector, d, with ndims(X) elements.

**length:**     Length of vector or largest array dimension

numberOfElements = length(array) finds the number of elements along the largest dimension of an array. array is an array of any MATLAB data type and any valid dimensions. numberOfElements is a whole number of the MATLAB double class.
For nonempty arrays, numberOfElements is equivalent to max(size(array)). For empty arrays, numberOfElements is zero.

**clf:**     Clear current figure window

clf deletes from the current figure all graphics objects whose handles are not hidden (i.e., their HandleVisibility property is set to on).

**\:**     The \-operator is used to solve a system of linear equations. If you are given the matrix $A$ and the right hand side $b$ of a system of linear equations

$$Ax = b$$

then the solution is computed in MATLAB with

```
>> x=A\b
```

**dot-operations:**     a dot preceding the operators *, /, ^ causes an element-by-element operation. Thus if $x$ and $y$ are vectors of the same length then

$$x. * y = [x_1 y_1, x_2 y_2, \ldots, x_n y_n].$$

**plot3:**     The plot3 function displays a three-dimensional plot of a set of data points.

**meshgrid:**     Rectangular grid in 2-D space.

[X,Y] = meshgrid(xgv,ygv) replicates the grid vectors xgv and ygv to produce a full grid. This grid is represented by the output coordinate arrays X and Y. The output coordinate arrays X and Y contain copies of the grid vectors xgv and ygv respectively. The sizes of the output arrays are determined by the length of the grid vectors. For grid vectors xgv and ygv of length M and N respectively, X and Y will have N rows and M columns.

**mesh:**  Mesh plot.

mesh(X,Y,Z) draws a wireframe mesh with color determined by Z, so color is proportional to surface height. If X and Y are vectors, length(X) = n and length(Y) = m, where [m,n] = size(Z). In this case, (X(j), Y(i), Z(i,j)) are the intersections of the wireframe grid lines; X and Y correspond to the columns and rows of Z, respectively. If X and Y are matrices, (X(i,j), Y(i,j), Z(i,j)) are the intersections of the wireframe grid lines.

**surf:**  3-D shaded surface plot.

surf(X,Y,Z) uses Z for the color data and surface height. X and Y are vectors or matrices defining the x and y components of a surface. If X and Y are vectors, length(X) = n and length(Y) = m, where [m,n] = size(Z). In this case, the vertices of the surface faces are (X(j), Y(i), Z(i,j)) triples. To create X and Y matrices for arbitrary domains, use the meshgrid function.

**contour:**  draws a contour plot of a matrix.

contour(Z) draws a contour plot of matrix Z, where Z is interpreted as heights with respect to the x-y plane. Z must be at least a 2-by-2 matrix that contains at least two different values. The x values correspond to the column indices of Z and the y values correspond to the row indices of Z. The contour levels are chosen automatically.
contour(Z,n) draws a contour plot of matrix Z with n contour levels where n is a scalar. The contour levels are chosen automatically.

**hold:**  Retain current graph when adding new graph

The hold function controls whether MATLAB clears the current graph when you make subsequent calls to plotting functions (the default), or adds a new graph to the current graph.

## 3.7 Problems

1. We are given the points

$$
\begin{array}{c|cccccc}
x & 0.9 & 2.3 & 3.9 & 4.6 & 5.8 & 7.3 \\
\hline
y & 2.9 & 4.1 & 4.8 & 7.0 & 7.0 & 8.7
\end{array}
$$

   (a) Define a region to plot the points using `axis`. Use `hold` to freeze the axis.
   (b) Plot the points using the symbol 'x'.
   (c) We want to fit a regression line through the points, that means compute the parameters $a$ and $b$ such that

$$
y_k = ax_k + b, \quad k = 1, \ldots, 6.
$$

   This is a linear system of equations with two unknowns and 6 equations. It cannot be solved exactly, the equations contradict themselves. However, the MATLAB \-operator does solve the system in the least squares sense by computing the best approximation for all equations.
   Form the linear system $A\binom{a}{b} = y$ and solve it by `A\y`.
   (d) Using the computed values of $a$ and $b$, plot the regression line on the same plot with the points.

2. Ellipse plots.

   (a) Plot the ellipse with center in origin and the main axis $a = 3$ on the $x$-axis and minor axis $b = 1$. Plot also the center using the symbol '+'.
   (b) Now move the ellipse so that the center is the point $(4, -1)$ and the direction of the main axis has an angle of $-30°$ with the $x$-axis. Plot this new ellipse in the same frame.

   Hint: Use a rotation matrix of the form

$$
Q = \begin{pmatrix} \cos\alpha & -\sin\alpha \\ \sin\alpha & \cos\alpha \end{pmatrix}
$$

   to rotate the coordinates of the ellipse.
3. Plot for $-3 \le x \le 3$ and $-5 \le y \le 5$ the function $f(x, y) = x^2 - 2yx^3$ using contour and mesh.

# Chapter 4
# Some Elementary Functions

Standard functions are available in almost all programming languages. But how could we actually compute them by using only the 4 basic arithmetic operations $\{+, -, \times, /\}$? Algorithms for computing standard functions on computers were developed some 60 years ago. Today many of them are part of the hardware. In spite of that it is interesting and challenging to try to develop a good algorithm for some well known functions.

A useful tool to evaluate a function is its Taylor expansion:

$$f(x) = f(a) + \frac{f'(a)}{1!}h + \cdots + \frac{f^{(n)}(a)}{n!}h^n + R_n(a, x), \qquad (4.1)$$

with $h = x - a$. For the remainder we have the estimate

$$|R_n(x, a)| \leq \frac{|f^{(n+1)}(\xi)|}{(n+1)!}|h|^{n+1}$$

where $\xi$ is a number between $a$ and $x$. If the function is infinitely differentiable then the remainder converges for $n \to \infty$ and for some $r$ (=the radius of convergence) for all $|x - a| < r$ to zero. Thus we get the Taylor series

$$f(x) = \sum_{i=0}^{\infty} \frac{f^{(i)}(a)}{i!}(x - a)^i, \qquad (4.2)$$

which represents the function for $|x - a| < r$. If the expansion point $a = 0$ then the series is called Maclaurin-series.

© Springer International Publishing Switzerland 2015
W. Gander, *Learning MATLAB*, UNITEXT - La Matematica per il 3+2 95,
DOI 10.1007/978-3-319-25327-5_4

## 4.1   Computing the Exponential Function

For $f(x) = e^x$ all derivatives exist $f^{(n)}(x) = e^x$ thus with $f^{(n)}(0) = 1$ the Maclaurin-series becomes

$$e^x = \sum_{i=0}^{\infty} \frac{x^i}{i!}. \tag{4.3}$$

One can show that in this case the radius of convergence is $r = \infty$, thus Eq. (4.3) can be used to compute $e^x$ for any $x$. Let's develop a program to sum up the series (4.3). The term

$$a_i = \frac{x^i}{i!}$$

can be computed by updating the preceding term $a_{i-1}$ through

$$a_i = \frac{x}{i} a_{i-1}.$$

We denote by sn the new and by so the old partial sum. We terminate the summation when the relative difference between the new and the old partial sum is smaller than $10^{-6}$. Thus we obtain a first version:

```
function sn=e1(x);
%
so=0; sn=1; term=1; k=1;
while abs(sn-so)>1e-6*sn
    so=sn; term=term*x/k;
    sn=so+term; k=k+1;
end
```

Indeed we obtain

```
>> for x=[1,-1,10,-10,20,-20]
     [e1(x) exp(x)]
   end
ans =
    2.718281801146385    2.718281828459046
ans =
    0.367879464285714    0.367879441171442
ans =
    1.0e+04 *
    2.202646026627129    2.202646579480672
ans =
    1.0e-04 *
    0.453999364851671    0.453999297624848
ans =
    1.0e+08 *
    4.851649751360876    4.851651954097902
ans =
    1.0e-08 *
    0.562188480727156    0.206115362243856
```

quite good results except for $x = -20$. We shall explain in the exercises why this
algorithm fails for large negative arguments. Since the algorithm seems to work well
for $x > 0$ we can make it work also for negative arguments by using the relation

$$e^{-x} = \frac{1}{e^x} \tag{4.4}$$

thus compute first $s = e^{|x|}$ and then adjust the result

```
if x<0, s=1/s; end
```

We can also improve the termination criterion. For fixed $x$ the terms

$$a_k = \frac{x^k}{k!} \to 0, \quad k \to \infty$$

converge rapidly to zero. So we shall sum the series until in finite arithmetic we get
for the partial sum $s_k$

$$s_k + a_{k+1} = s_k.$$

Thus we get the algorithm

```
function sn=Exp(x);
% EXP stable computation of the exponential function
%    s=Exp(x); computes an approximation s of exp(x) up to machine
%    precision.

if x<0, v=-1; x=abs(x); else v=1; end
so=0; sn=1; term=1; k=1;
while sn~=so
    so=sn; term=term*x/k;
    sn=so+term; k=k+1;
end
if v<0, sn=1/sn; end;
```

This program works now perfectly

```
>> for x=[1,-1,10,-10,20,-20]
      (exp(x)-Exp(x))/exp(x)
   end
ans =
      0
ans =
   1.5089e-16
ans =
   3.3033e-16
ans =
  -2.9851e-16
ans =
   1.2285e-16
ans =
  -2.0066e-16
```

We obtain results that match the exponential function to machine precision.

## 4.2   Computing sin and cos

The Taylor series for the two trigonometric functions are obtained by the beautiful
relation

$$e^{ix} = \cos x + i \sin x \quad \text{Euler's formula} \tag{4.5}$$

by expanding the Maclaurin series for $e^{ix}$ and separating real and imaginary parts:

$$\cos x = \sum_{k=0}^{\infty} \frac{(-1)^k}{(2k)!} x^{2k} = 1 - \frac{x^2}{2!} + \frac{x^4}{4!} - \cdots \tag{4.6}$$

$$\sin x = \sum_{k=0}^{\infty} \frac{(-1)^k}{(2k+1)!} x^{2k+1} = x - \frac{x^3}{3!} + \frac{x^5}{5!} - \cdots \tag{4.7}$$

If we wish to use these expansions for computing $\cos x$ or $\sin x$ we have to expect
numerical problems for large arguments $|x| \gg 1$. This because the series are alter-
nating and affected by cancellation. The remedy is to *reduce* first the argument to the
interval $[-\pi/2, \pi/2]$. Then the series can be summed without too much cancellation
(see Problem 2).

## 4.3   Computing arctan

We know that the derivative of $f(x) = \arctan x$ is given by

$$\frac{d}{dx} \arctan x = \frac{1}{1+x^2} = \sum_{k=0}^{\infty} (-1)^k x^{2k}$$

The series converges for $|x| < 1$. Thus by integrating we obtain

$$\arctan x = \sum_{k=0}^{\infty} (-1)^k \frac{x^{2k+1}}{2k+1} = x - \frac{x^3}{3} + \frac{x^5}{5} - \cdots \tag{4.8}$$

We shall use this series to compute arctan in Problem 5.

## 4.4   MATLAB-Elements Used in This Chapter

**for:**   Execute statements specified number of times.
for index=values, program statements, end
repeatedly executes one or more MATLAB statements in a loop.
values can be:

- initval:endval
  increments the index variable from initval to endval by 1, and repeats execution of program statements until index is greater than endval.
- initval:step:endval
  increments index by the value step on each iteration, or decrements when step is negative.
- valArray
  The loop executes for a maximum of n times, where n is the number of columns of valArray.

**while:** Repeatedly execute statements while condition is true.
while expression, statements, end
repeatedly executes one or more MATLAB program statements in a loop as long as an expression remains true

**mod:** Modulus after division
M = mod(X,Y) returns the modulus after division of X by Y. In general, if Y does not equal 0, M = mod(X,Y) returns X - n.*Y, where n = floor(X./Y). If Y is not an integer and the quotient X./Y is within roundoff error of an integer, then n is that integer. Inputs X and Y must have the same dimensions unless one is a scalar double. If one input has an integer data type, then the other input must be of the same integer data type or be a scalar double. (will be used in the problem section)

## 4.5 Problems

1. Explain what happens in Algorithm e1 when $x = -20$. *Hint:* look at the size of the largest term and at the final result. What happens when computing the result in finite arithmetic?
2. Write a MATLAB-function to compute $\sin x$ using the series (4.7). In order to avoid cancellation for large $|x|$ reduce the argument to the interval $[-\pi/2, \pi/2]$.
3. Do the same for $\cos x$.
4. Combine both functions and write a function to compute $\tan x$.
5. Write a function to compute $\arctan x$ for $|x| < 1$ using the series (4.8) and compare your result with the standard MATLAB-function atan(x).

# Chapter 5
# Computing with Multiple Precision

In this section we shall show how to perform some computations with more digits than given by the IEEE floating point arithmetic. The problems of this section will need integer operations and variables of integer data types. It is an opportunity to learn the corresponding MATLAB features.

## 5.1 Computation of the Euler Number $e$

We shall compute the Euler number $e = \exp(1)$ to an arbitrary number of decimal digits. For this we will use the algorithm e1 in Chap. 4 which we developed to compute the exponential function using the Taylor series. The series is evaluated for for $x = 1$:

$$e = \sum_{k=0}^{\infty} \frac{1}{k!}. \tag{5.1}$$

Using the notations $a = 1/k!$ for the $k$th term and $s$ for a partial sum we get the function

```
function s=Eulerconstant;
s=1;sn=2; a=1; k=1;    % initialization
while s~=sn
  s=sn; k=k+1;
  a=a/k;               % new term
  sn=s+a;              % new partial sum
end
```

© Springer International Publishing Switzerland 2015
W. Gander, *Learning MATLAB*, UNITEXT - La Matematica per il 3+2 95,
DOI 10.1007/978-3-319-25327-5_5

Indeed we obtain with

```
>> format long
>> s=Eulerconstant
s =
   2.718281828459046
```

a result with 16 decimal digits which is what we can expect by using IEEE-arithmetic. If we want to compute more digits we need to simulate multiple precision arithmetic. In the above algorithm the only arithmetic operations are

$$k = k + 1, \quad a = a/k \quad \text{and} \quad sn = s + a.$$

For $k$ we may use a simple variable. We shall not compute so many terms of the Taylor series that also $k$ has to be represented in multiple precision arithmetic. The partial sum and the terms to be added, however, have to be *multiple precision numbers*. We shall store the digits of a multiple precision number in a integer array. There are several integer data types in MATLAB available. Here we shall use `uint32`[1]. which is a data type for unsigned integers.

   This function is used for conversion to 32-bit unsigned integers. These 32-bit numbers cover the range from 0 to $2^{32} - 1 = 4294967295$. Let $a$ be an array of such unsigned integer numbers. We represent the number 1 using 20 digits:

```
>> a=uint32(zeros(1,20));
>> a(1)=1
a =
  Columns 1 through 10
     1     0     0     0     0     0     0     0     0     0
  Columns 11 through 20
     0     0     0     0     0     0     0     0     0     0
```

If we wish to divide this number by `k=2` we should get the result

```
a =
  Columns 1 through 10
     0     5     0     0     0     0     0     0     0     0
  Columns 11 through 20
     0     0     0     0     0     0     0     0     0     0
```

Another division by `k=3` should give

```
a =
  Columns 1 through 10
     0     1     6     6     6     6     6     6     6     6
  Columns 11 through 20
     6     6     6     6     6     6     6     6     6     6
```

---

[1]For Numeric MATLAB Types see http://www.mathworks.com/help/matlab/numeric-types.html

We would like to see the digits not as elements of a vector but continuously as large number. This can be done by using the function `sprintf` (`sprintf` formats data into a string of characters).

```
>> sprintf('%01d',a)
ans =
01666666666666666666
```

Let us program this division. We need to use a function for integer division. In MATLAB this is the function `idivide`. For the remainder we use the function `mod`. Suppose we want to divide 14 by 3. The result is: `quotient = 14/3 = 4` and `remainder = mod(14,3) = 2`. Programmed in MATLAB this is

```
quotient=idivide(14,3)   remainder=mod(14,3)
```

Thus our division function becomes

```
function a=Divide1(k,a)
% divides the integer array a by integer number k
n=length(a);
c=10;
remainder=a(1);
for i=1:n-1
  a(i)=idivide(remainder,k);
  remainder=mod(remainder,k)*c+a(i+1);
end
a(n)=idivide(remainder,k);
```

Let us test this function. The following script divides the initial number `a=1` with the numbers $k = 2, \ldots, 15$. Thus the last result should be $1/15!$:

```
clear, clc, format long
res=[];
a=zeros(1,30,'uint32');
a(1)=1;
for k=2:15
  a=Divide1(k,a);
  res =[res;sprintf('%01d',a)];
end
res
1/factorial(15)

res =
050000000000000000000000000000
016666666666666666666666666666
004166666666666666666666666666
000833333333333333333333333333
000138888888888888888888888888
000019841269841269841269841269
000002480158730158730158730158
```

```
00000027557319223985890652557 3
00000027557319223985890652557
00000002505210838544171877505
00000000208767569878680989792
00000000016059043836821614599
00000000001147074559772972471
0000000000007647163731819816 4
>> 1/factorial(15)
ans =
     7.647163731819816e-13
```

It looks good! Notice that the smaller the numbers get the more leading zeros appear.
It is not necessary to divide these leading zeros by $k$ since they remain zero. We use
the variable imin to count the number of leading zeros in the vector and start the
division at position a(imin+1). The division function changes so to

```
function [A,imin]=Divide(c,imin,k,A)
% DIVISION divides the multiple precision number A by the integer
% number k.  The first imin components of A are zero. imin is updated
% after the division. c defines the number of decimal digits in one
% array element: c=10 is used for one digit, c=100 for two digits etc.
n=length(A);
if imin <n                            % if imin=n => A=0
  first=1;
  remainder=A(imin+1);
  for i=imin+1:n
    A(i)=idivide(remainder,k);
    if A(i)==0
      if first                       % update imin
        imin=i;
      end
    else
      first=0;
    end
    if i<n
      remainder=mod(remainder,k)*c+A(i+1);
    end
  end
end
```

Notice at the beginning of the division the variable imin is updated if a(i) becomes
zero. After the first nonzero element the update is stopped. The following test shows
that imin counts the leading zeros correctly:

```
% Testprogram for Divide for c=10 or c=100
clear, clc, format long
a=zeros(1,30,'uint32');
imin=0;
a(1)=1;
c=100;
if c==10, w='%01d'; else w='%02d'; end
```

```
res =[sprintf('%5d',imin),'    ',sprintf(w,a)];
for k=2:15
  [a,imin]=Divide(c,imin,k,a);
  res =[res; sprintf('%5d',imin),'    ',sprintf(w,a)];
end
res


res =
    0    100000000000000000000000000000000
    1    050000000000000000000000000000000
    1    016666666666666666666666666666666
    2    004166666666666666666666666666666
    3    000833333333333333333333333333333
    3    000138888888888888888888888888888
    4    000019841269841269841269841269841269
    5    000002480158730158730158730158
    6    000000275573192239858906525573
    7    000000027557319223985890652557
    8    000000002505210838544171877505
    9    000000000208767569878680989792
   10    000000000016059043836821614599
   11    000000000001147074559772972471
   13    000000000000076471637318198164
```

So far we can generate the terms of the series of Eq. (5.1). Next we need to sum up the terms. Let s denote the partial sum and t the next term. The function Add.m is straightforward

```
function r=Add(imin,s,a);
% ADD adds the multiprecision number a to s without carry. It is
% supposed that s>a and that imin leading components of a are zero
n=length(s);
r=s;
for i=imin+1:n
  r(i)=s(i)+a(i);
end
```

but we have to take care of possible carry and so we need also to sweep over the array with

```
function s=Carry(c,s);
% CARRY normalizes the component of s such that 0 <= s(i) < c
% and moves the carry to the next component
n=length(s);
for i=n:-1:2
  while s(i)>=c
    s(i)=s(i)-c; s(i-1)=s(i-1)+1;
  end
end
```

Our main program becomes

```
function s=EmultPrec(c,n);
% EMULTPREC computes n*log10(c) decimal digits of
% the Euler constant e=exp(1). The digits are stored
% in the array s.
a=zeros(1,n,'uint32');        % define array of
s=a;                          % unsigned integers
a(1)=1; s(1)=2;
k=1; imin=0;                  % imin skips leading zeros
while imin<n
  k=k+1;
  [a,imin]=Divide(c,imin,k,a);  % new term
  s=Add(imin,s,a);              % new partial sum
  s=Carry(c,s);
end
```

With these preparations we can now compute

```
>> s=EmultPrec(10,10)
s =
    2    7    1    8    2    8    1    8    2    3
>> e=sprintf('%01d',s)
e =
2718281823
```

To compute more digits we use

```
>> s=EmultPrec(10,60); e=sprintf('%01d',s)
e =
271828182845904523536028747135266249775724709369995957496673
```

Because of Rounding errors some of the last printed digits are not correct. We can
check this by computing 10 more digits:

```
>> s=EmultPrec(10,70); e=sprintf('%01d',s)
e =
2718281828459045235360287471352662497757247093699959574966967627724050
```

So we see that the last two digits of s=EmultPrec(10,60) are affected by
rounding errors.

**Packing More Digits in One Array Element**

The parameter $c$ of EmultPrec controls how many digits are stored in one array
element. If we change it to c=100 we work with two digits per array element and
get

```
>> s=EmultPrec(100,30); e=sprintf('%01d',s)
e =
271828182845945235362874713526624977572479369999595749646
```

Note that the printed result is wrong! Zeros are missing, e.g. for the sequence after the 13th digit we get 5945 instead of 59045. We have to adjust the format to print 2 digits with leading zero if necessary:

```
>> s=EmultPrec(100,30); e=sprintf('%02d',s)
e =
0271828182845904523536028747135266249775724709369999595749646
```

Now the result is correct.

## 5.2   MATLAB-Elements Used in This Chapter

**uint32:**   Convert to 32-bit unsigned integer.
intArray = uint32(array) converts the elements of an array into unsigned 32-bit (4-byte) integers of class uint32.
intArray: Array of class uint32. Values range from 0 to $2^{32} - 1$

**zeros:**   Create array of all zeros.
X = zeros(sz) returns an array of zeros where size vector sz defines size(X). For example, zeros([2 3]) returns a 2-by-3 matrix.
X = zeros(1,3,'uint32')
Create a 1-by-3 vector of zeros whose elements are 32-bit unsigned integers.

**idivide:**   Integer division with rounding option.
C = idivide(A, B) is the same as A./B except that fractional quotients are rounded toward zero to the nearest integers.

**mod:**   Modulus after division.
M = mod(X,Y) returns the modulus after division of X by Y. In general, if Y does not equal 0, M = mod(X,Y) returns X − n.*Y, where n = floor(X./Y).

**sprintf:**   Format data into string
str = sprintf(formatSpec,A1,...,An) formats the data in arrays A1,...,An according to formatSpec in column order, and returns the results to string str.

## 5.3   Problems

For the following problems, make use of the functions we developed for computing Euler's number *e*.

1. Compute using multiple precision the powers of 2:

$$2^i, \quad i = 1, 2, \ldots, 300.$$

2. Write a program to compute factorials using multiple precision:

$$n!, \quad n = 1, 2, \ldots, 200.$$

3. Compute $\pi$ to 1000 decimal digits. Use the relation by C. Størmer:

$$\pi = 24 \arctan \frac{1}{8} + 8 \arctan \frac{1}{57} + 4 \arctan \frac{1}{239}.$$

**Hints**:

- Compute first a multiprecision arctan function using the Taylor-series (4.8) as proposed in Chap. 4:

$$\arctan x = \sum_{k=0}^{\infty} (-1)^k \frac{x^{2k+1}}{2k+1} = x - \frac{x^3}{3} + \frac{x^5}{5} - \cdots$$

- The above series is alternating so there is a danger of cancellation. However, since it is used only for $|x| < 1$ this is not much a concern. What we need is a new funtion Sub

```
function r=Sub(c,a,b)
% SUB computes r=a-b where a and b are multiprecision numbers
% with a>b.
```

  to subtract two multiprecision numbers. One has to be careful not to generate negative numbers, all intermediate results have to remain positive.
- To compute $\pi$ we have to evaluate for some integer $p > 1$ the function $\arctan(1/p)$. When generating the next term after

$$t_k = \frac{x^{2k+1}}{2k+1}$$

for $x = 1/p$ we have to form

$$t_{k+1} = t_k/p^2/(2k+1).$$

There is bug that one has to avoid: by dividing the last term twice by $p$ and a third time by $2k + 1$ the variable imin is updated. For the next term we need to know the value of imin before the division by $2k + 1$! Otherwise we will get erroneous results when forming $t_k/p^2$.

# Chapter 6
# Solving Linear Equations

In this chapter we shall develop a method to solve linear equations. Given a matrix $A$ and a right hand side $b$

$$A = \begin{pmatrix} a_{11} & a_{12} & \cdots & a_{1n} \\ \vdots & \vdots & \cdots & \vdots \\ a_{m1} & a_{m2} & \cdots & a_{mn} \end{pmatrix}, \quad b = \begin{pmatrix} b_1 \\ \vdots \\ b_m \end{pmatrix}$$

then a solution of $Ax = b$ is given in MATLAB by x=A\b. Notice that MATLAB produces a solution for any values of $n$ and $m$. The \-operator is very powerful but to understand it one has to know what it computes. We shall explain here only the cases when $m = n$ and $m > n$.

$m = n$: We have as many equations as unknowns. If the matrix $A$ is non-singular then the linear system has exactly one solution $x = A^{-1}b$.

$m > n$: The system is overdetermined, it has more equations than unknowns. In general there is no solution, the equations contradict themselves. A way to find a good "solution" was given by Carl Friedrich Gauss who invented the *Least Squares Method*. This method determines a solution vector $x$ by minimizing the length of the *residual* $r = b - Ax$:

$$\|r\|_2^2 = \|b - Ax\|_2^2 = \min.$$

## 6.1 Gaussian Elimination and $LU$ Decomposition

Given an $n \times n$ linear system, the usual way to compute a solution is by Gaussian elimination. In the first step the unknown $x_1$ is eliminated from equations two to $n$, leaving a reduced $(n-1) \times (n-1)$-system containing only the unknowns $x_2, \ldots, x_n$. Continuing the elimination steps we finally obtain an equation with the only unknown $x_n$. The original system is thus transformed and reduced to a triangular system $Ux = y$:

© Springer International Publishing Switzerland 2015
W. Gander, *Learning MATLAB*, UNITEXT - La Matematica per il 3+2 95,
DOI 10.1007/978-3-319-25327-5_6

$$\begin{pmatrix} u_{11} & u_{12} & \cdots & u_{1n} \\ & u_{22} & \cdots & u_{2n} \\ & & \ddots & \vdots \\ & & & u_{nn} \end{pmatrix} \begin{pmatrix} x_1 \\ x_2 \\ \vdots \\ x_n \end{pmatrix} = \begin{pmatrix} y_1 \\ y_2 \\ \vdots \\ y_n \end{pmatrix}$$

and the solution is easily computed by *back-substitution*. One can show that by this elimination process the matrix $A$ is factorized in a product of a unit lower triangular matrix $L$ and an upper triangular matrix $U$:

$$A = LU.$$

The elimination process is mathematically equivalent with transforming the linear system by multiplying it from the left with the non-singular matrix $L^{-1}$:

$$Ax = b \implies L^{-1}Ax = L^{-1}b \iff Ux = y.$$

Things get a bit more complicated if the equation which should be used for eliminating $x_k$ does not contain this variable. Then one has to permute the equations to continue the elimination process. This *partial pivoting strategy* is used when solving equations by the \-operator. If no equation is found which contains $x_k$ in the $k$th elimination step or if the coefficient of $x_k$ is very small then the system is considered to be singular and a warning message is issued:

```
Warning: Matrix is close to singular or badly scaled. Results
  may be inaccurate. RCOND =  4.800964e-18.
```

The MATLAB-function `lu` computes the LU-factorization of a matrix. We enter a $5 \times 5$ matrix:

```
>> A = [12,1,2,2,10; 14 4 15 6 1
         2  8  14  14  13
        14 14 7 12 14
         9 14 12 14 10]
A =
    12     1     2     2    10
    14     4    15     6     1
     2     8    14    14    13
    14    14     7    12    14
     9    14    12    14    10
```

Notice that the elements on a row are separated by spaces or by a comma. A new row can either be started on the same line by inserting a semicolon or by typing it on a new line. Next we call the function `lu`:

```
>> [L,U,P]=lu(A)
L =
       1.0000            0            0            0            0
       0.6429       1.0000            0            0            0
       0.8571      -0.2125       1.0000            0            0
       0.1429       0.6500      -0.9970       1.0000            0
       1.0000       0.8750       0.9716      -0.3442       1.0000
U =
      14.0000       4.0000      15.0000       6.0000       1.0000
            0      11.4286       2.3571      10.1429       9.3571
            0            0     -10.3562      -0.9875      11.1312
            0            0            0       5.5655      17.8727
            0            0            0            0       0.1483
P =
       0       1       0       0       0
       0       0       0       0       1
       1       0       0       0       0
       0       0       1       0       0
       0       0       0       1       0
```

We obtain two triangular matrices $L$ and $U$ and a permutation matrix $P$. They are related by

$$PA = LU.$$

Let's check this and compute $LU$ and $PA$:

```
>> L*U
ans =
      14.0000       4.0000      15.0000       6.0000       1.0000
       9.0000      14.0000      12.0000      14.0000      10.0000
      12.0000       1.0000       2.0000       2.0000      10.0000
       2.0000       8.0000      14.0000      14.0000      13.0000
      14.0000      14.0000       7.0000      12.0000      14.0000
>> P*A
ans =
      14       4      15       6       1
       9      14      12      14      10
      12       1       2       2      10
       2       8      14      14      13
      14      14       7      12      14
```

As expected, the product $LU$ is equal to the permuted matrix $A$.

Consider now a vector for the right hand side:

```
>> b=[1:5]'
b =
     1
     2
     3
     4
     5
```

Notice that $1:5$ is the abbreviation for the vector $[1,2,3,4,5]$ and the apostrophe transposes the vector to become a column vector. The solution of $Ax = b$ is given by x=A\b.

```
>> x=A\b
x =
      5.4751
    -13.7880
    -11.1287
     25.9130
     -8.0482
```

The \-operator computes in this case first a $LU$-decomposition and then obtains the solution by solving $Ly = Pb$ by forward-substitution followed by solving $Ux = y$ with backward-substitution. We can check this with the following statements.

```
>> y=P*b
y =
     2
     5
     1
     3
     4
>> y=L\y
y =
      2.0000
      3.7143
      0.0750
      0.3748
     -1.1939
>> x=U\y
x =
      5.4751
    -13.7880
    -11.1287
     25.9130
     -8.0482
```

## 6.2 Elimination with Givens-Rotations

In this section we present another elimination algorithm which is computationally more expensive but simpler to program and which can be used also for least squares problem.

We proceed as follow to eliminate in the $i$th step the unknown $x_i$ in equations $i + 1$ to $n$. Let

$$(i) : a_{ii}x_i + \ldots + a_{in}x_n = b_i$$
$$\vdots \qquad \qquad \vdots$$
$$(k) : a_{ki}x_i + \ldots + a_{kn}x_n = b_k \qquad (6.1)$$
$$\vdots \qquad \qquad \vdots$$
$$(n) : a_{ni}x_i + \ldots + a_{nn}x_n = b_n$$

be the reduced system. To eliminate $x_i$ in equation $(k)$ we multiply equation $(i)$ by $-\sin \alpha$ and equation $(k)$ by $\cos \alpha$ and replace equation $(k)$ by the linear combination

$$(k)_{new} = -\sin \alpha \cdot (i) + \cos \alpha \cdot (k), \qquad (6.2)$$

where we have chosen $\alpha$ so, that

$$a_{ki}^{new} = -\sin \alpha \cdot a_{ii} + \cos \alpha \cdot a_{ki} = 0. \qquad (6.3)$$

No elimination is necessary if $a_{ki} = 0$, otherwise we can use Eq. (6.3) to compute

$$\cot \alpha = \frac{a_{ii}}{a_{ki}} \qquad (6.4)$$

and get

$$
\begin{aligned}
&\cot = A(i, i)/A(k, i); \\
&\text{si } = 1/sqrt(1 + \cot * \cot); \\
&\text{co } = \text{si} * \cot.
\end{aligned}
\qquad (6.5)
$$

In this elimination we do not only replace equation $(k)$ but seemingly unnecessarily also equation $(i)$ by

$$(i)_{new} = \cos \alpha \cdot (i) + \sin \alpha \cdot (k). \qquad (6.6)$$

Doing so we do not need to permute the equations as with Gaussian Elimination. This is done automatically. We illustrate this for the case if $a_{ii} = 0$ and $a_{ki} \neq 0$. Here we obtain $\cot \alpha = 0$ thus $\sin \alpha = 1$ and $\cos \alpha = 0$. Computing the two Eqs. (6.2) and (6.6) results in just permuting them!

The Givens Elimination algorithm is easy to program since we can use MATLAB's vector-operations. To multiply the $i$th row of the matrix $A$ by a factor `co`$= \cos(\alpha)$

$$\cos(\alpha)[a_{i1}, a_{i,2}, \ldots, a_{in}]$$

we use the statement

```
co*A(i,:)
```

The colon notation is an abbreviation for `A(i,1:n)` or more general `A(i,1:end)`. The variable `end` serves as the last index in an indexing expression. Thus the new $i$th row of the matrix becomes

```
A(i,:)=co*A(i,:)+si*A(k,:)
```

By doing so we overwrite the $i$th row of $A$ with new elements. This makes it impossible to compute the new $k$th row since we need the old values of the $i$th row! We have to save the new row first in a auxiliary variable h and assign it later:

```
A(i,i)=A(i,i)*co+A(k,i)*si;
h=A(i,i+1:n)*co+A(k,i+1:n)*si;
A(k,i+1:n)=-A(i,i+1:n)*si+A(k,i+1:n)*co;
A(i,i+1:n)=h;
```

Since `A(k,i)` becomes zero we do not compute it. Also we do not use `A(i,:)` but rather `A(i,i+1:n)` since the elements on the row before the diagonal are zero and don't have to be processed.

We propose here a more elegant solution without auxiliary variable. The Givens elimination is performed by transforming the two rows with a rotation matrix

$$\begin{pmatrix} c & s \\ -s & c \end{pmatrix} \begin{pmatrix} A(i,i) & A(i,i+1) & \ldots & A(i,n) \\ A(k,i) & A(k,i+1) & \ldots & A(k,n) \end{pmatrix}$$

An assignment statement in MATLAB cannot have two results. But by using the expression `A(i:k-i:k,i+1:n)` we can change both rows of $A$ in one assignment with one result. Thus we get the compact assignments

```
A(i,i)=A(i,i)*co+A(k,i)*si;
S=[co,si;-si,co];
A(i:k-i:k,i+1:n)=S*A(i:k-i:k,i+1:n);
```

In the same way we also change the right hand side. Putting all together we obtain the function:

```
function x=EliminationGivens(A,b);
% ELIMINATIONGIVENS solves a linear system using Givens-rotations
%    x=EliminationGivens(A,b) solves Ax=b using Givens-rotations.
[m,n]=size(A);
for i= 1:n
  for k=i+1:m
    if A(k,i)~=0
      cot=A(i,i)/A(k,i);                        % rotation angle
      si=1/sqrt(1+cot^2); co=si*cot;
      A(i,i)=A(i,i)*co+A(k,i)*si;
      S=[co,si;-si,co];
      A(i:k-i:k,i+1:n)=S*A(i:k-i:k,i+1:n);
      b(i:k-i:k)=S*b(i:k-i:k);
    end
  end;
  if A(i,i)==0
    error('Matrix is rank deficient');
  end;
end
x=zeros(n,1);
for k=n:-1:1          % backsubstitution
  x(k)=(b(k)-A(k,k+1:n)*x(k+1:n))/A(k,k);
end
x=x(:);
```

The transformation of $A x = b$ to the reduced system $U x = y$ is done this time with Givens rotations. These rotation matrices are not only non-singular but also *orthogonal* (a matrix $Q$ is orthogonal if $Q^\top Q = I$). Transformations with orthogonal matrices leave the length invariant:

$$z = Q r \implies \|z\|_2^2 = z^\top z = (Q r)^\top Q r = r^\top Q^\top Q r = r^\top r = \|r\|_2^2.$$

Therefore the solution of minimizing the length of the residual $r = b - A x$ does not change of we multiply the system by an orthogonal matrix:

$$A x = b \iff Q^\top A x = Q^\top b.$$

With Givens elimination we therefore can solve linear $n \times n$-systems and also overdetermined systems in the least square sense. The MATLAB \-operator solves least squares systems using orthogonal transformations.

We illustrate this with the following example. We wish to fit a function of the form

$$y = a t + \frac{b}{t} + c \sqrt{t}$$

to the points

$$\begin{array}{c|ccccc} t & 1 & 2 & 3 & 4 & 5 \\ \hline y & 2.1 & 1.6 & 1.9 & 2.5 & 3.1 \end{array}$$

Inserting the points we get the linear system

$$\begin{pmatrix} 1 & 1 & 1 \\ 2 & 1/2 & \sqrt{2} \\ 3 & 1/3 & \sqrt{3} \\ 4 & 1/4 & \sqrt{4} \\ 5 & 1/5 & \sqrt{5} \end{pmatrix} \begin{pmatrix} a \\ b \\ c \end{pmatrix} = \begin{pmatrix} 2.1 \\ 1.6 \\ 1.9 \\ 2.5 \\ 3.1 \end{pmatrix}$$

We get with GivensElimination the same result as with MATLAB's \-operator:

```
% file CurveFit.m
A=[ 1    1    1
    2    1/2   sqrt(2)
    3    1/3   sqrt(3)
    4    1/4   sqrt(4)
    5    1/5   sqrt(5)];
b=[ 2.1 1.6 1.9 2.5 3.1]';

t=[1:5]';
plot(t,b,'o')
hold
x=EliminationGivens(A,b)
y=A\b
pause          % compare x with y, they are the same
xx=[1:0.1:5];
yy=y(1).*xx+y(2)./xx+y(3).*sqrt(xx);
plot(xx,yy)


>> CurveFit
x =
     1.0040
     2.1367
    -1.0424
y =
     1.0040
     2.1367
    -1.0424
```

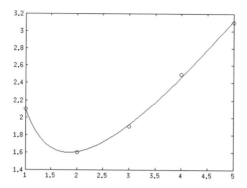

## 6.3    MATLAB-Elements Used in This Chapter

**mldivide**, \ :  (backslash-operator) Solve systems of linear equations Ax = B for x

x = A\B solves the system of linear equations A*x = B. The matrices A and B must have the same number of rows. MATLAB displays a warning message if A is badly scaled or nearly singular, but performs the calculation regardless.

**: Colon**

```
J:K   is the same as [J, J+1, ..., K].
J:K   is empty if J>K.
J:D:K  is the same as [J, J+D, ..., J+m*D] where m=fix((K-J)/D).
J:D:K  is empty if D == 0, if D>0 and J>K, or if D<0 and J<K.

COLON(J,K) is the same as J:K and COLON(J,D,K) is the same as J:D:K.

The colon notation can be used to pick out selected rows, columns
and elements of vectors, matrices, and arrays.  A(:) is all the
elements of A, regarded as a single column. On the left side of an
assignment statement, A(:) fills A, preserving its shape from before.
A(:,J) is the J-th column of A.  A(J:K) is [A(J),A(J+1),...,A(K)].
A(:,J:K) is [A(:,J),A(:,J+1),...,A(:,K)] and so on.
```

**lu:**

[L,U,P] = LU(A) returns unit lower triangular matrix L, upper triangular matrix U, and permutation matrix P so that P*A = L*U.

**planerot:**  Givens plane rotation

[G,y] = planerot(x) where x is a 2-component column vector, returns a 2-by-2 orthogonal matrix G so that y = G*x has y(2) = 0.

**' ctranspose:**  Complex conjugate transpose

b = a' computes the complex conjugate transpose of matrix a and returns the result in b.
The following commands are used in the problem section.

**pause:**            Halt execution temporarily

pause, by itself, causes the currently executing function to stop and wait for you to press any key before continuing. Pausing must be enabled for this to take effect. (See pause on, below.) Pause without arguments also blocks execution of Simulink models, but not repainting of them.

pause(n) pauses execution for n seconds before continuing, where n is any nonnegative real number. Pausing must be enabled for this to take effect.

**tril:**             Lower triangular part of matrix

L = tril(X) returns the lower triangular part of X.

L = tril(X,k) returns the elements on and below the kth diagonal of X.
k = 0 is the main diagonal, k > 0 is above the main diagonal, and k < 0 is below the main diagonal.

**triu:**             Upper triangular part of matrix

U = triu(X) returns the upper triangular part of X.

U = triu(X,k) returns the element on and above the kth diagonal of X.
k = 0 is the main diagonal, k > 0 is above the main diagonal, and k < 0 is below the main diagonal.

**diag:**             Create diagonal matrix or get diagonal elements of matrix

D = diag(v) returns a square diagonal matrix with the elements of vector v on the main diagonal.

D = diag(v,k) places the elements of vector v on the kth diagonal. k = 0 represents the main diagonal, k > 0 is above the main diagonal, and k < 0 is below the main diagonal.

x = diag(A) returns a column vector of the main diagonal elements of A.

x = diag(A,k) returns a column vector of the elements on the kth diagonal of A.

## 6.4  Problems

1. LU-decomposition Consider the linear system $Ax = b$ defined by the matrix

```
>> format short e, format compact
>> n=5; A=invhilb(n), b=eye(n,1)
```

(a) Apply Gaussian Elimination (without pivoting) to reduce the system to
$U x = y$

```
for j=1:n-1                          % Elimination
  for k=j+1:n
    fak=A(k,j)/A(j,j);
    A(k,j:n)=A(k,j:n)-fak*A(j,j:n);
    b(k)=b(k)-fak*b(j);
  end
end
```

Watch the elimination process by displaying the matrix and the right hand
side after each elimination step. Use the pause statement to stop execution.
(b) Next store the factors fak instead of the zeros you introduce by eliminating $x_j$:

```
for j=1:n-1                          % Elimination
  for k=j+1:n
    fak=A(k,j)/A(j,j);
    A(k,j)=fak;                      % store factors instead zeros
    A(k,j+1:n)=A(k,j+1:n)-fak*A(j,j+1:n);
  end
end
```

Now use the commands triu, tril, diag to extract $L$ and $U$ from $A$ and
verify that indeed $LU = A$.

2. Replace the computation of the rotation matrix S in our function
EliminationGivens by the MATLAB-function planerot. Convince your-
self that you get the same results with the modified function by solving the curve
fitting example again.
3. Determine the parameters $a$ and $b$ such that the function $f(x) = ae^{bx}$ fits the
following data

| $x$ | 30.0 | 64.5 | 74.5 | 86.7 | 94.5 | 98.9 |
|-----|------|------|------|------|------|------|
| $y$ | 4    | 18   | 29   | 51   | 73   | 90   |

Plot the points and the fitted function.

**Hint:** If you fit log $f(x)$ the problem becomes very easy!

4. The following statistics lists the population of Shanghai since 1953:

| year | in million |
|------|-----------|
| 1953 | 6.2044    |
| 1964 | 10.8165   |
| 1982 | 11.8597   |
| 1990 | 13.3419   |
| 2000 | 16.4077   |
| 2010 | 23.0192   |

Fit a polynomial through these data and predict the population for 2016 and 2020.
Plot your results.

5. *Fitting of circles.* We are given the measured points $(\xi_i, \eta_i)$:

| $\xi$ | 0.7 | 3.3 | 5.6 | 7.5 | 6.4 | 4.4 | 0.3 | −1.1 |
|---|---|---|---|---|---|---|---|---|
| $\eta$ | 4.0 | 4.7 | 4.0 | 1.3 | −1.1 | −3.0 | −2.5 | 1.3 |

Find the center $(c_1, c_2)$ and the radius $r$ of a circle $(x - c_1)^2 + (y - c_2)^2 = r^2$ that
approximate the points as well as possible. Consider the *algebraic fit*: Rearrange
the equation of the circle as

$$2c_1 x + 2c_2 y + r^2 - c_1^2 - c_2^2 = x^2 + y^2. \tag{6.7}$$

With $w = r^2 - c_1^2 - c_2^2$, we obtain with (6.7) for each measured point a linear
equation for the unknowns $c_1$, $c_2$ and $w$.

- Write a function function drawcircle(C,r) to plot a circle with center (C(1),C(2)) and radius r.
- Computer the center and the radius and plot the given points and the fitted
  circle.

6. Seven dwarfs are sitting around a table. Each one has a cup. The cups contain milk,
   all together a total of 3 liter. One of the dwarfs starts distributing his milk evenly
   over all cups. After he has finished his right neighbor does the same. Clockwise
   the next dwarfs proceed distributing their milk. After the 7th dwarf has distributed
   his milk, there is in each cup as much milk as at the beginning. How much milk
   was initially in each cup?

   **Hint:** Let $x = (x_1, x_2, \ldots, x_7)^\top$ be the initial milk distribution. Thus $\sum_{j=1}^{7} x_j = 3$. Simulate the distributing of milk as matrix-vector Operation:

$$x^{(1)} = T_1 x.$$

After 7 distributions you obtain $x^{(7)} = x$ and thus

$$x = T_7 T_6 \cdots T_1 x$$

or $(A - I)x = 0$ where $A = T_7 T_6 \cdots T_1$. Add to this homogeneous sys-
tem the equation $\sum_{j=1}^{7} x_j = 3$ and solve the system using our function
EliminationGivens. Compare the results you get with those when using
MATLAB's \-operator.

7. The following sections were measured on the street $\overline{AD}$ depicted in Fig. 6.1.

$$AD = 89\,\text{m}, \ AC = 67\,\text{m}, \ BD = 53\,\text{m}, \ AB = 35\,\text{m and } CD = 20\,\text{m}$$

Balance out the measured sections using the least squares method.

**Fig. 6.1**  Street

# Chapter 7
# Recursion

## 7.1  Introduction

Recursion is a powerful concept in computer science. The basic idea is that the solution of a problem often can be reduced to solving some smaller instances of the same problem. Recursive solutions can be applied to many problems, one well known strategy is called *divide and conquer*.

A function is sometimes defined recursively. For instance the factorial, the function

$$f(n) = n! = 1 \times 2 \times 3 \times \cdots \times n$$

can be defined recursively by

$$0! = 1$$
$$f(n) = n \times f(n-1), \quad n > 0.$$

In MATLAB we could just use the expression `prod(1:n)` or `factorial(n)` to compute this function. To show the concept of recursion we program the function

```
function f=Factorial(n)
if n==0,
    f=1;
else
    f=n*Factorial(n-1);
end
```

The function `f` is calling itself in its definition. This is called a recursive function. A recursion that contains only one single self-reference is known as *single recursion*, while a recursion that contains multiple self-references is known as *multiple recursion*. A single recursion can be programmed easily as an iteration, which is simpler and more efficient. For our factorial example we would get

© Springer International Publishing Switzerland 2015
W. Gander, *Learning MATLAB*, UNITEXT - La Matematica per il 3+2 95,
DOI 10.1007/978-3-319-25327-5_7

```
function f=FactorialIteration(n)
f=1;
for k=1:n
  f=k^f;
end
```

We get for all variants the same

```
>> n=10;
>> [prod(1:n),factorial(n), FactorialIteration(n), Factorial(n)]
ans =
     3628800      3628800      3628800      3628800
```

## 7.2  Laplace Expansion for Determinants

The determinant of a matrix $A$ can be computed using the *Laplace Expansion*. For each row $i$ we have

$$\det(A) = \sum_{j=1}^{n} a_{ij}(-1)^{i+j}\det(M_{ij}), \tag{7.1}$$

where $M_{ij}$ denotes the $(n-1) \times (n-1)$ submatrix obtained by deleting row $i$ and column $j$ of the matrix $A$. Thus computing the determinant of $A$ is reduced to compute $n$ smaller determinants of the submatrices $M_{ij}$. Instead of expanding the determinant as in (7.1) along a row, we can also use an expansion along a column.

The following MATLAB program is an example for multiple recursion, it computes a determinant using the Laplace Expansion for the first row:

```
function d=DetLaplace(A);
% DETLAPLACE determinant using Laplace expansion
%    d=DetLaplace(A); computes the determinant d of the matrix A
%    using the Laplace expansion for the first row.

n=length(A);
if n==1;
  d=A(1,1);
else
  d=0; v=1;
  for j=1:n
    M1j=[A(2:n,1:j-1) A(2:n,j+1:n)];
    d=d+v*A(1,j)*DetLaplace(M1j);
    v=-v;
  end
end
```

In MATLAB the function det computes the determinant in a more efficient way (using Gaussian Elimination) than our recursive function. The following examples show both results:

```
for n=4:9
  A=rand(n);
  [det(A) DetLaplace(A)]
end
ans =
   0.128257928707307      0.128257928707307
ans =
  -0.084250098064663     -0.084250098064664
ans =
  -0.181256419130385     -0.181256419130385
ans =
  -0.022309977397375     -0.022309977397376
ans =
  -0.006338537112776     -0.006338537112776
ans =
  -0.008692776468285     -0.008692776468285
```

The results are the same, the only difference is that Laplace's formula needs much more operations and thus uses much more execution time that det. However, if we replace rand by hilb, the matrices are Hilbert matrices which are ill-conditioned, we get

```
for n=4:9
  A=hilb(n);
  [det(A) DetLaplace(A)]
end
ans =
   1.0e-06 *
   0.165343915343926      0.165343915343319
ans =
   1.0e-11 *
   0.374929513251423      0.374929513075645
ans =
   1.0e-17 *
   0.536729988684877      0.536730023323187
ans =
   1.0e-24 *
   0.483580261909806      0.483085292821939
ans =
   1.0e-30 *
   0.002737050274535     -0.355714248182654
ans =
   1.0e-35 *
   0.000000097202790      0.315086992638140
```

This time the results of det are much more accurate than those with Laplace's Expansion. Thus in summary, Laplace's expansion is mathematically interesting and can be implemented recursively. It is, however, computationally much more expensive and numerically a disaster.

## 7.3   Hilbert Curves

Hilbert curves are space filling curves. We follow here the derivation given in [13]. The basic elements for the construction are 4 "cups".

      a                   b                   c                   d

The next refinement of $a$ is the curve $a_2$ which is

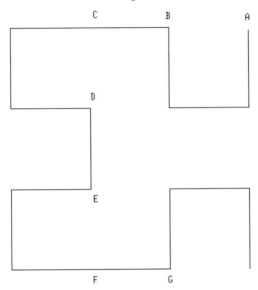

It has been constructed by attaching a smaller $d$-cup to point $A$, the upper right of $a$, then moving a step $h$ from point $B$ to $C$ to the left to place a small $a$-cup. Then moving down from point $D$ to $E$ again by a step $h$ and placing another $a$-cup. Finally moving from $F$ to $G$ another step $h$ to the right and placing a $b$-cup. We wish to plot the resulting Hilbert curve, so we need to plot the segments $\overline{BC}$, $\overline{DE}$ and $\overline{FG}$. We thus get the function

```
function a(i);
global x y h;
if i>0,
    d(i-1); plot([x-h,x],[y,y]); x=x-h;
    a(i-1); plot([x,x],[y-h,y]); y=y-h;
    a(i-1); plot([x,x+h],[y,y]); x=x+h;
    b(i-1);
end
```

The above curve is the result for $i = 2$. Symbolically we write for the construction

$$a_2 : d \leftarrow a \downarrow a \rightarrow b$$

The coordinates of the current point on the curve are given by the global variables $(x, y)$. The current step size $h$ is also a global variable.

Similarly we get for the $d$-cup starting from the upper right corner:

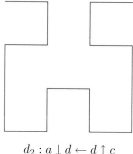

$$d_2 : a \downarrow d \leftarrow d \uparrow c$$

An finally for $b$ and $c$ starting from the lower left corner:

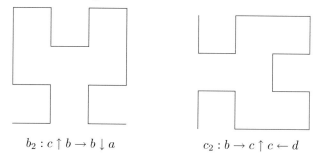

$$b_2 : c \uparrow b \rightarrow b \downarrow a \qquad\qquad c_2 : b \rightarrow c \uparrow c \leftarrow d$$

We program the 4 cases:

```
function a(i);
global x y h;
if i>0,
d(i-1); plot([x-h,x],[y,y]); x=x-h;
a(i-1); plot([x,x],[y-h,y]); y=y-h;
a(i-1); plot([x,x+h],[y,y]); x=x+h;
b(i-1);
end

function b(i);
global x y h;
if i>0,
c(i-1); plot([x,x],[y,y+h]); y=y+h;
b(i-1); plot([x,x+h],[y,y]); x=x+h;
b(i-1); plot([x,x],[y-h,y]); y=y-h;
a(i-1);
end
```

```
function c(i);
global x y h;
if i>0,
  b(i-1); plot([x,x+h],[y,y]); x=x+h;
  c(i-1); plot([x,x],[y,y+h]); y=y+h;
  c(i-1); plot([x-h,x],[y,y]); x=x-h;
  d(i-1);
end

function d(i);
global x y h;
if i>0,
  a(i-1); plot([x,x],[y-h,y]); y=y-h;
  d(i-1); plot([x-h,x],[y,y]); x=x-h;
  d(i-1); plot([x,x],[y,y+h]); y=y+h;
  c(i-1);
end
```

Notice that these four functions are highly multiple recursive. Each one calls itself
and two of the other functions.

In the main program h1.m we need to define *n* and call a(n):

```
global x y h;
clf
axis([-600,800, -600, 800])
axis square
hold
x=600; y=600 ;
n=input('Hilbert Curve n=?')
h0=1024;
h=h0/2^n;                   % scaling to fill same square
a(n)
```

The result for $n = 3$ is

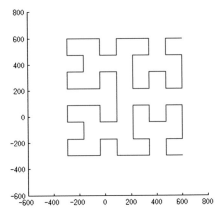

It is fun to watch the curve being plotted. For this we need to add the MATLAB-
command drawnow while the curve is generated. We add this command to the
function a.m:

```
function a(i);
global x y h
if i>0,
  d(i-1); plot([x-h,x],[y,y]); x=x-h;
  a(i-1); plot([x,x],[y-h,y]); y=y-h;
  a(i-1); plot([x,x+h],[y,y]); x=x+h;
  b(i-1);
 drawnow;
end
```

Run now the main program again for $n = 6$ and watch the curve being plotted!

## 7.4  Quicksort

Quicksort is a ingenious sorting algorithm developed by Tony Hoare using recursion. Given a vector of numbers

$$\boldsymbol{a} = (a_1, a_2, \ldots, a_n)$$

we want to sort them in ascending order. We split the vector in two sets by choosing some element in the middle:

$$x = a(m) \quad \text{where} \quad m = \text{round}((1 + n)/2)$$

We then get two sets which are separated by $x$:

$$\{a_1, \ldots, a_{m-1}\}, x, \{a_{m+1}, \ldots, a_n\}.$$

Now we scan the elements of the first set and search for an element $a_i \geq x$. Then we scan the elements of the second set and look for an element $a_j \leq x$. If we are successful, we swap $a_i$ with $a_j$. We continue this way until the first set contains only numbers smaller than $x$ and the second set only numbers larger than $x$. Then we apply the same procedure recursively and independently to the two sets.

The following function `Sorting` generates first $n$ random numbers and prints them as bar plot.

```
function Sorting(n)
global a
format short
a=rand(1,n);
clf, bar(a), pause
quick(1,n)
a
```

Then it calls the recursive function `quick`:

```
function quick(left,right)
% QUICK quicksort
% quick(left,right) sorts the numbers a(left), ..., a(right) of the
%         global array a in ascending order.
global a;
mid=round((left+right)/2);        % choose middle element
i=left; j=right; x=a(mid);        % sort a(i) ... a(j)
while i<=j
  while a(i)<x, i=i+1; end         % search left for  a(i)>=x
  while x<a(j), j=j-1; end         % search right for a(j)<=x
  if i<=j                          % swap if found
    u=a(i); a(i)=a(j); a(j)=u;
    i=i+1; j=j-1;                  % advance indices
%      bar(a); pause(0.01)          % to show the process
  end
end
if left<j, quick(left,j) ; end    % sort the two sets
if i<right,quick(i,right); end    % recursively
```

When two elements are swapped we plot the array $a$ and wait 0.01 s. This allows to visualize the quick-sort algorithm. Run the program for $n = 100$ and watch how the numbers are sorted. In a parallel environment the sorting of the two subsets could be computed independently.

## 7.5   MATLAB-Elements Used in This Chapter

**prod**:          Product of array elements

                   If A is a vector, then prod(A) returns the product of the elements.
                   If A is a nonempty matrix, then prod(A) treats the columns of A as
                   vectors and returns a row vector of the products of each column.

**factorial**:     Factorial of input

                   f = factorial(n) returns the product of all positive integers less than or
                   equal to n, where n is a nonnegative integer value. If n is an array, then
                   f contains the factorial of each value of n. The data type and size of f
                   is the same as that of n.

**det**:           Matrix determinant

                   d = det(X) returns the determinant of the square matrix X.

**rand**:          Uniformly distributed pseudorandom numbers

                   r = rand returns a pseudorandom scalar drawn from the standard uni-
                   form distribution on the open interval (0, 1).

**hilb**:          Hilbert matrix

                   The Hilbert matrix is a notable example of a poorly conditioned matrix.
                   The elements of the Hilbert matrices are H(i,j) = 1/(i + j 1).

**global**:        Declare global variables

                   global X Y Z defines X, Y, and Z as global in scope.

                   Ordinarily, each MATLAB function has its own local variables, which
                   are separate from those of other functions, and from those of the
                   base workspace. However, if several functions, and possibly the base
                   workspace, all declare a particular name as global, they all share a
                   single copy of that variable. Any assignment to that variable, in any
                   function, is available to all the functions declaring it global.

**drawnow**:       Update figure window and execute pending callbacks drawnow causes

                   figure windows and their children to update. Any callbacks generated
                   by user actions (for example, mouse or key presses, button clicks, and
                   so on) are executed before drawnow returns.

Use drawnow in animation loops to update the figure during function execution and to update graphical user interfaces.

**bar:** Bar graph

bar(Y) draws one bar for each element in Y.

**tic, toc:** stopwatch timer

tic starts a stopwatch timer to measure performance. The function records the internal time at execution of the tic command. Display the elapsed time with the toc function.

## 7.6 Problems

1. Cramer's Rule for solving systems of linear equations. This rule is often used when solving small ($n \leq 3$) systems of linear equations by hand.
   Write a function x=Cramer(A,b) which solves a linear system $Ax = b$ using Cramer's rule. For $\det(A) \neq 0$, the linear system has the unique solution

$$x_i = \frac{\det(A_i)}{\det(A)}, \quad i = 1, 2, \ldots, n, \tag{7.2}$$

   where $A_i$ is the matrix obtained from $A$ by replacing column $a_{:i}$ by $b$. Use the function DetLaplace to compute the determinants.
   Test your program by generating a linear system with known solution.

2. Selection Sort versus Quick Sort.
   The idea of selection sort is to find the minimum value in the given array and then swaps it with the value in the first position. By repeating this for the remaining elements the array is sorted.

   (a) Write a (non-recursive) function a=SelectSort(a) which implements the Selection Sort. Show the process using bar and pause as done in Quick Sort. Test your program by sorting some small arrays ($n \leq 100$).
   (b) Speed Test: Remove the bar and pause statement in both functions and measure the time each function needs to sort an array of 100,000 elements. Use for this the MATLAB-functions tic and toc.
   (c) For fun (not efficient!): program the selection sort recursively. Use a global array and proceed similarly as with quicksort.

3. Pythagoras Tree[1]:
   Basic construction: Given two points $P$ and $Q$ in the plane, construct the points $P'$ and $Q'$ to built a square. Then put on the square a right triangle with one basis angle $\alpha$.

---

[1] https://en.wikipedia.org/wiki/Pythagoras_tree_(fractal).

The following figure shows the basic construction and the first recursion step, where the construction is repeated on top of the cathetes of the triangle $\overline{P'RQ'}$.

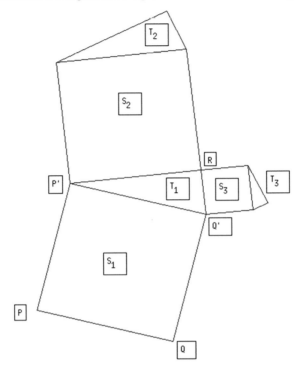

Write a recursive function which computes the Pythagoras tree until the base line $\overline{PQ}$ becomes small. Experiment with the basis angle, choose e.g. as here in the figure $\alpha = 20°$.

# Chapter 8
# Iteration and Nonlinear Equations

## 8.1 Bisection

Consider the following problem: We are given the area $F = 12$ of a right-angled triangle and the section $p = 2$ of the hypotenuse (see Fig. 8.1). Compute the edges of the triangle.

Denote with $q$ the second section of the hypotenuse so that $c = p + q$. The height theorem says $h_c^2 = pq$. Replacing $c$ and $h_c$ in the expression for the area $F = \frac{1}{2}ch_c$ we get

$$F = \frac{p+q}{2}\sqrt{pq}$$

which is an interesting relation since it says that the area of the triangle is equal to the product of the arithmetic and the geometric mean of the two sections of the hypotenuse. Inserting the numerical values we get an equation for $x = q$:

$$f(x) = \frac{2+x}{2}\sqrt{2x} - 12 = 0.$$

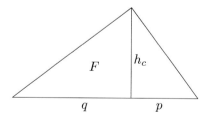

**Fig. 8.1** Triangle problem

© Springer International Publishing Switzerland 2015
W. Gander, *Learning MATLAB*, UNITEXT - La Matematica per il 3+2 95,
DOI 10.1007/978-3-319-25327-5_8

We wish to find a value $x$ such that $f(x) = 0$. It is easy to see that $f(0) = -12$ and $f(8) = 8$, so we conclude, since $f$ is a continuous function in $(0, 8)$, that there exists a zero in this interval.

It is obvious to try the mid-point of the interval $x = \frac{0+8}{2} = 4$. We obtain $f(4) = 3\sqrt{8} - 12 = -3.5147$ and conclude that our solution must be in the smaller interval $(4, 8)$. Another bisection $x = \frac{4+8}{2} = 6$ gives $f(6) = 1.8564$ thus the solution must be in the interval $(4, 6)$. We can continue this process called *bisection* until the two bounds are close enough to give us the solution to the precision we wish to have.

The following function `Bisekt` can be used for this:

```
function x=Bisekt(f,a,b)
x=(a+b)/2;
while b-a>1e-5
  if f(x)<0,
      a=x;
  else
      b=x;
  end
  x=(a+b)/2
end
```

Indeed we get the solution with

```
>> x=Bisekt(@(x)(2+x)/2*sqrt(2*x)-12,4,6)
x =
    5.3423
```

Note that the function `Bisekt` must be improved to serve as a more general root-finder. We will do that in Problem 1.

## 8.2   Newton's Method

Let $s$ be a simple zero of the function $f$. We want to compute $s$ by approximating $f$ by a simpler function $h(x)$ near $s$. The solution $x_1$ of $h(x) = 0$ is then an approximation of $s$. Newton's method approximates $f$ by the Taylor-polynomial of degree one at $x_0$ in the neighborhood of $s$

$$h(x) = f(x_0) + f'(x_0)(x - x_0).$$

The solution of $h(x) = 0$ is

$$x_1 = x_0 - \frac{f(x_0)}{f'(x_0)}.$$

This is called a Newton iteration step. Repeating the computation generates a sequence $\{x_k\}$ which usually converges to the solution $s$.

### 8.2.1 Algorithm of Heron

As example consider the function $f(x) = x^2 - a$, where $a > 0$. The positive solution of $f(x) = 0$ is $s = \sqrt{a}$. Applying Newton's method we get the iteration

$$x_{k+1} = x_k - \frac{f(x_k)}{f'(x_k)} = x_k - \frac{x_k^2 - a}{2x_k} = \frac{1}{2}\left(x_k + \frac{a}{x_k}\right).$$

The sequence generated by this iteration is a method to compute the square-root using only the four basic operations. It is known as the *Algorithm of Heron*. In Problem 7 a careful implementation is discussed.

### 8.2.2 Fractal

Consider the function $f(z) = z^3 - 1$. It has the three roots, one is real the other two are complex:

$$z_1 = 1, \quad z_2 = -\frac{1}{2} + \frac{\sqrt{3}}{2}i, \quad z_3 = -\frac{1}{2} - \frac{\sqrt{3}}{2}i.$$

Using Newton's iteration

$$z_{k+1} = z_k - \frac{z_k^3 - 1}{3z_k^2}$$

and complex arithmetic, the sequence $\{z_k\}$ will converge to one of the roots depending of the starting value $z_0$. It is interesting to determine which starting point leads to which root. The set of initial values that lead to convergence to the same root is called the *basin of attraction* of that root. As we will see the basins of attraction of the roots have a very complicated structure, and similarly for their boundaries: they are *fractal*.

We consider the region in the complex plane $\{z = x + iy \mid -1 \le x, y \le 1\}$. We will choose 1,000 points in each direction as starting values for Newton's iteration. Thus a million points will be used. In the following program script we make use of two features of MATLAB: vector-operations and meshgrid. The points are generated with the function meshgrid and stored as complex numbers in the matrix Z. The iteration is performed in parallel with the whole matrix Z.

```
clf,clc,clear
n=1000; m=30;
x=-1:2/n:1;
[X,Y]=meshgrid(x,x);
Z=X+1i*Y;                 % define grid for picture
for i=1:m                 % perform m iterations in parallel
   Z=Z-(Z.^3-1)./(3*Z.^2);  % for all million points
end;                      % if converged then
```

```
                              % each element of Z contains one root
a=20;                         % transform roots to pos. integer values
image((round(imag(Z))+2)*a);    % multiply by a to get nice colors
```

After $m = 30$ iterations each element of the matrix Z has converged to one of the roots. To interpret Z as an image, we need to transform the elements to real numbers. We can distinguish the three different elements by looking at their imaginary part which is 0, $\frac{\sqrt{3}}{2}$, or $-\frac{\sqrt{3}}{2}$. By rounding and adding 2 we get the numbers 2, 3, and 1. Now to choose nice colors we multiply them by a factor $a$. For $a = 20$ we obtain the picture

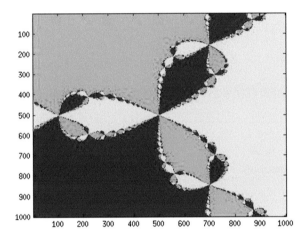

## 8.3   Circular Billiard

We consider a *circular billiard table* and two balls located at the points $P$ and $Q$, see Fig. 8.2. In which direction must the ball at point $P$ be hit, if it is to bounce off the boundary of the table exactly once and then hit the other ball located at $Q$?

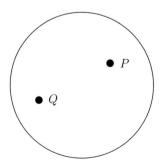

**Fig. 8.2**  Billiard table

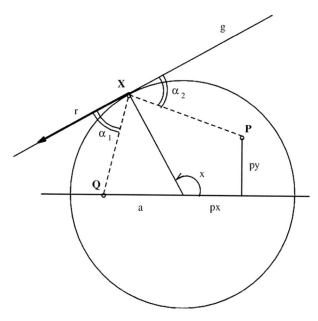

**Fig. 8.3** Billiard problem

The problem does not depend on the size of the circle. Therefore, without loss of generality, we may assume the radius of the table to be 1, i.e., we will consider the unit circle. Also, the problem remains the same if we rotate the table. Thus, we may assume that one ball (e.g. $Q$) is located on the $x$-axis.

The problem can now be stated as follows: In the unit circle, two arbitrary points $P = (p_x, p_y)$ and $Q = (a, 0)$ are given. We are looking for a reflection point $X = (\cos x, \sin x)$ (see Fig. 8.3) on the circumference of the circle, such that a billiard ball traveling from $P$ to $X$ will hit $Q$ after it bounces off the edge. The problem is solved if we know the point $X$, which means that we are looking for the angle $x$.

The condition that must be satisfied is that the two reflection angles are equal, i.e., $\alpha_1 = \alpha_2$ in Fig. 8.3. This is the case if the point $X$ is the bisector of the angle $QXP$. Thus if, $e_{XQ}$ is the unit vector in the direction $XQ$, and if $e_{XP}$ is defined similarly, then the direction of the bisector is given by the sum $e_{XQ} + e_{XP}$. This vector must be orthogonal to the direction vector of the tangent $g$,

$$r = \begin{pmatrix} \sin x \\ -\cos x \end{pmatrix}.$$

Therefore we obtain for the angle $x$ the equation

$$f(x) = (e_{XQ} + e_{XP})^\top r = 0. \tag{8.1}$$

Let us program now the function $f(x)$ which we will call `billiard(x)`.

```
function y=billiard(x)
% computes the billiard function f
global px py a
c=cos(x); s=sin(x);
X=[c;s]; P=[px;py];
XP=P-X; Ep=XP/norm(XP);        % unit vector direction XP
XQ=[a-c; -s]; Eq=XQ/norm(XQ);   % unit vector direction XQ
r=[s;-c];                       % tangent direction vector
y=(Ep+Eq)'*r;
```

As an example we plot the function for the following ball positions

$$P = (0.6, 0.3), \quad Q = (-0.2, 0)$$

```
% PlotBilliardFct.m
clear,clc,clf
global px py a
px=0.6,py=0.3,a=-0.2
F=[];
X=0:0.01:2*pi;
for x=X
  F=[F,billiard(x)];
end
plot(X,F,[0,2*pi],[0,0])
```

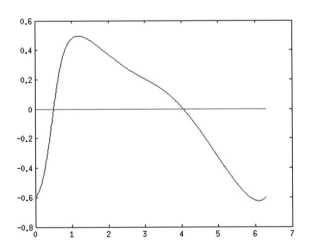

From this plot we see that there are two solutions for angles: one in the interval $(0, 1)$ and the other one in $(4, 5)$. Using bisection we get

```
>> [x1,y]=Bisection(@billiard,0,1)
x =
   0.504063160498908
y =
     2.220446049250313e-16
>> [x2,y]=Bisection(@billiard,4,5)
x =
   4.050212021055064
y =
    -2.220446049250313e-16
```

the reflection points

```
format short
>> X1=[cos(x1),sin(x1)]                    % first Reflection point
X1 =
    0.8756     0.4830
>> X2=[cos(x2),sin(x2)]                    % second Reflection point
X2 =
   -0.6148    -0.7887
```

Now we would like to plot the trajectories:

```
% BilliardExample2.m
clf
axis equal, hold
t=0:0.01:2*pi;                   % plot circle
plot(cos(t),sin(t))
plot(px,py,'o')                   % plot point P
text(px,py,'   P')
plot(a,0,'o')                     % plot point Q
text(a,0,'   Q')
P=[px,py]; Q=[a,0];
plot(X1(1),X1(2),'o')
text(X1(1),X1(2),'   X_1')
plot([Q(1),X1(1)], [Q(2),X1(2)])  % plot trajectory
plot([X1(1),P(1)], [X1(2),P(2)])
plot(X2(1),X2(2),'o')
text(X2(1),X2(2),'   X_2')
plot([Q(1),X2(1)], [Q(2),X2(2)])  % plot trajectory
plot([X2(1),P(1)], [X2(2),P(2)])
```

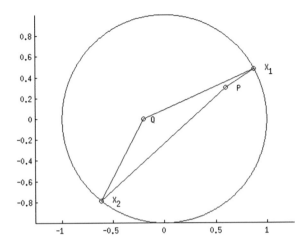

## 8.4   MATLAB-Elements Used in This Chapter

**meshgrid**:   Rectangular grid in 2-D and 3-D space
[X,Y] = meshgrid(xgv,ygv) replicates the grid vectors xgv and ygv to produce a full grid. This grid is represented by the output coordinate arrays X and Y. The output coordinate arrays X and Y contain copies of the grid vectors xgv and ygv respectively. The sizes of the output arrays are determined by the length of the grid vectors. For grid vectors xgv and ygv of length M and N respectively, X and Y will have N rows and M columns.

**imag**:   Imaginary part of complex number
imag(z) returns the imaginary part of z.
imag(A) returns the imaginary part of each element of A.

**image**:   Display image object
image creates an image graphics object by interpreting each element in a matrix as an index into the figure's colormap or directly as RGB values, depending on the data specified.

**text**:   Create text object in current axes
text is the low-level function for creating text graphics objects. Use text to place character strings at specified locations.
text(x,y,'string') adds the string in quotes to the location specified by the point (x,y) x and y must be numbers of class double.

## 8.5 Problems

1. Bisection-Algorithm. Improve the function `Bisekt`. Your function `[x,y]= Bisection(f,a,b,tol)` should also compute a zero for functions with $f(a) > 0$ and $f(b) < 0$ to a given tolerance `tol`. Be careful to stop the iteration in case the user asks for a too small tolerance! If by the bisection process we arrive at an interval $(a, b)$ which does not contain a machine number anymore then it is high time to stop the iteration.

2. Solve with bisection the equations

$$(a) \quad x^x = 50 \qquad (b) \quad \ln(x) = \cos(x) \qquad (c) \quad x + e^x = 0.$$

   **Hint**: a starting interval is easy to find by sketching the functions involved.

3. Find $x$ such that

$$\int_0^x e^{-t^2} dt = 0.5.$$

   **Hint**: the integral cannot be evaluated analytically, so expand it in a series and integrate. Write a function `f(x)` to evaluate the series. Then use bisection to compute the solution of $f(x) - 0.5 = 0$.

4. Use bisection to create the following table:

| F | 0 | $0.1\pi$ | $0.2\pi$ | ... | $\pi$ |
|---|---|---|---|---|---|
| h | 0 | ? | ? | ... | 2 |

   where the function $F(h)$ is given by

$$F(h) = \pi - 2\arccos\frac{h}{2} + h\sqrt{1 - \left(\frac{h}{2}\right)^2}.$$

5. *Binary search:* we are given an ordered sequence of numbers:

$$x_1 \le x_2 \le \cdots \le x_n$$

   and a new number $z$. Write a program that computes an index value $i$ such that either $x_{i-1} < z \le x_i$ or $i = 1$ or $i = n + 1$ holds. The problem can be solved by considering the function

$$f(i) = x_i - z$$

   and computing its "zero" by bisection.

6. Compute $x$ where the following maximum is attained:

$$\max_{0<x<\frac{\pi}{2}} \left( \frac{1}{4\sin x} + \frac{\sin x}{2x} - \frac{\cos x}{4x} \right).$$

7. Write a function s=SquareRoot(a) which computes the square root using Heron's algorithm. Think of a good starting value and a good termination criterion.
   **Hint**: consider the geometrical interpretation of Newton's method and use the (theoretical) monotonicity of the sequence as termination criterion.
   Test your function and compare the results with the standard MATLAB-function sqrt. Compute the relative error of both functions.
8. We consider again Problem 3: find $x$ such that

$$f(x) = \int_0^x e^{-t^2} dt - 0.5 = 0.$$

Since a function evaluation is expensive (summation of the Taylor series) but the derivatives are cheap to compute, a higher order method is appropriate. Solve this equation with Newton's method.

9. Using Newton's iteration, find $a$ such that $\int_0^1 e^{at} dt = 2$.

10. Consider the billiard-problem. Let the ball $P$ be at position $P = (0.5, 0.5)$ and let $Q$ move in small steps (say 0.1) from 1 to $-1$.
    Compute for each position the solutions using bisection. Count and plot the solutions and plot also the function billiard. make a pause before moving on the the next position of $Q$.
11. Modify the fractal program by replacing $f(z) = z^3 - 1$ with the function

$$f(z) = z^5 - 1.$$

   (a) Compute the 5 zeros of $f$ using the command roots.
   (b) In order two distinguish the 5 different numbers, study the imaginary parts of the 5 zeros. Invent a transformation such that the zeros are replaced by 5 different positive integer numbers.

12. Mandelbrot set[1]: Consider the iteration

$$Z_{k+1} = Z_k^2 + C.$$

Depending on the value of the constant $C$ the sequence $\{Z_k\}$ will either diverge to $\pm\infty$ or converge.

---

[1] This problem is nicely solved and discussed in [8].

Let $C$ now be in the region in the complex plane $Z = X + iY$ with $-2 \leq X, Y \leq 2$.

Perform 50 iterations starting always with $Z_0 = 0$ with all numbers $C$ in that region and plot using `image` the resulting Mandelbrot set, which is the set of all values $C$ for which the iterations converges to a finite limit.

# Chapter 9
# Simulation

## 9.1  Event Simulation Using Random Numbers

In this section we investigate experimental—and theoretical probabilities. The experimental probability is the quotient of the number of times the event occurs divided by the total number of trials. The theoretical probability on the other hand is the number of favorable outcomes divided by the total number of possible outcomes.

The following example can be solved analytically and by simulation. Consider the following experiment: we are tossing a coin $n = 10$ times. Since for a fair coin the theoretical probability for head or tail is $1/2$ we expect in our experiment that $k \approx 5$ times head to appear. In fact by calling the function toss a few times we obtain what we expect:

```
function k=toss(n)
k=sum(rand(1,n)<=0.5);

>> k=toss(10)
k =      5
>> k=toss(10)
k =      6
>> k=toss(10)
k =      5
>> k=toss(10)
k =      4
>> k=toss(10)
k =      7
>> k=toss(10)
k =      5
>> k=toss(10)
k =      6
```

It is interesting to repeat the experiment, say $m = 100$ times and to make a histogram of the numbers $k$ obtained (see Fig. 9.1).

© Springer International Publishing Switzerland 2015
W. Gander, *Learning MATLAB*, UNITEXT - La Matematica per il 3+2 95,
DOI 10.1007/978-3-319-25327-5_9

**Fig. 9.1** Histogram

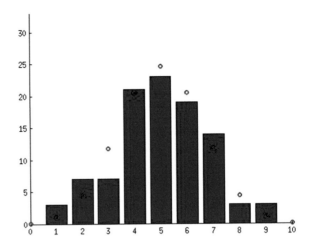

```
% khist1.m
% Flipping a coin, histogram
clear, clf
format compact
m=100
n=10
a=zeros(1,n);                    % array for histogram
z=0;
for p=1:m
  k=toss(n);
     if k==0                     % exception because Matlab
        z=z+1;                   % does not allow to use a(0)
     else
        a(k)=a(k)+1;             % update array a
     end
end
axis([0,n,0,max(a)+n])
hold
bar([0:n],[z,a])
```

Since MATLAB does not a allows zero indices we treat zero as a special case.

Let us now compute the theoretical probability. The number of ways to get $k$ heads in $n$ flips is given by the binomial coefficient ("$n$ choose $k$"):

$$\binom{n}{k}.$$

For the theoretical probability we have to divide by the total number of possible outcomes which is

$$\frac{\binom{n}{k}}{\sum_{j=0}^{n}\binom{n}{j}} = \frac{\binom{n}{k}}{2^n}.$$

Notice that the sum $\sum_{j=0}^{n} \binom{n}{j} = (1+1)^n = 2^n$.

To compute this we first need the function

```
function b=binomial(n,k)
b=1;
for j=1:k
  b=b*(n+1-j)/j;
end
```

Then the theoretical probablity is computed by the statements

```
for k=0:n
  if k==0
    z=binomial(n,k);
  else
  f(k)=binomial(n,k);        % theoretical probability
  end
end
f=f/2^n;z=z/2^n;             % probability
z=z*m;  f=f*m;               % scaled by m
plot([0:n],[z,f],'or','LineWidth',2)
```

This function is shown using red circles in Fig. 9.1, it has a "bell shape" which is known as normal distribution.

By increasing $m$, the histogram and the theoretical probability of a given number of heads becomes smoother and approaches as limit the normal distribution:

$$\frac{1}{\sqrt{2\pi}\sigma} \exp\left(-\frac{(k-\mu)^2}{2\sigma^2}\right)$$

where $\mu = n/2$ is the mean and $\sigma$ the standard deviation, a measure of the breadth of the curve width. We program this normal distribution function as

```
function y=bell(x,mu,sigma)
y=1/sqrt(2*pi)/sigma*exp(-(x-mu).^2/2/sigma^2);
```

For equal probability coin flipping we have $\sigma = \sqrt{\mu/2} = \sqrt{n/4}$. For $n = 10$ and $m = 100$ we get with

```
% khist2.m
figure(2)
axis([0,n,0,max(a)+n])
hold
plot([0:n],[z,f],'or','LineWidth',2)
sigma=sqrt(n/4)
mu=n/2
x=linspace(0,n);
plot(x,bell(x,mu,sigma)*m,'LineWidth',2)
```

Figure 9.2 which shows already a very good match between the discrete distribution and the continuous asymptotic normal distribution.

Consider now a larger simulation with $n = 100$ and $m = 10'000$. We get with

**Fig. 9.2** Discrete and
normal distribution

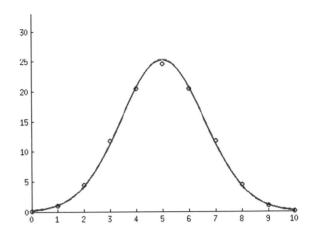

```
% khist.m
clear, clf
format compact
m=10000
n=100
a=zeros(1,n);               % array for histogram
z=0;
for p=1:m
  k=toss(n);
    if k==0                 % Matlab allows no a(0)
       z=z+1;
    else
       a(k)=a(k)+1;         % update array a
    end
end
axis([0,n,0,max(a)+n])
hold
bar([0:n],[z,a])
sigmaT=sqrt(n/4)
meanT=n/2
x=linspace(0,n);
plot(x,bell(x,meanT,sigmaT)*m,'r','LineWidth',2)

m =
        10000
n =
     100
Current plot held
sigmaT =   5
meanT =   50
```

We notice first from Fig. 9.3 that the histogram matches very well the normal distribution and that it is rare that $k < 40$ or $k > 60$. In fact summing up 95 % of all cases
around $k = 50$ we get

**Fig. 9.3**  Khist

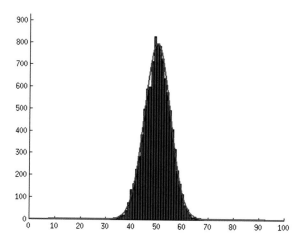

```
s=a(n/2);
p=0;
while s<=m*95/100
  p=p+1;
  s=s+a(n/2-p)+a(n/2+p);
end
p


p = 10
```

which means that only in 5 % of all cases $k$ was outside the interval $(40, 60)$ which
is the interval `meanT` $\pm 2 \times$ `sigmaT`.

If we do not know the theoretical mean and the standard deviation then for our
histogram these two quantities can be estimated by

$$\mu = \frac{1}{n} \sum_{i=1}^{n} a_i x_i, \quad \sigma = \sqrt{\frac{1}{n} \sum_{i=1}^{n} a_i (x_i - \mu)^2}.$$

For our example we obtain

```
meanEx= sum(a.*[1:n])/m
deviations=[1:n]-meanEx;
variance=sum(a.*deviations.^2)/m
sigmaEx=sqrt(variance)


meanEx =
   49.9325
variance =
   25.2867
sigmaEx =
    5.0286
```

which compares very well with the theoretical quantities.

## 9.2   Exhaustive Search

Exhaustive search is a problem solving technique to find an optimum in some finite space by enumerating and inspecting all possible states.

These types of algorithms are of limited use since enumerating all permutations of $n$ objects means inspecting a matrix with $n!$ rows and $n$ columns. The MATLAB-function `perms` computes all permutations of $n$ objects. In the description of this function we find the warning:

> This function is only practical for situations where N is less than about 10 (for N=11, the output takes over 3 gigabytes).

It would be better not to generate and store all permutations at once but produce and use each permutation sequentially.

On the other hand, our computers have become very powerful regarding processing time and memory. So for small but nevertheless interesting problems exhaustive search can be a valuable technique.

As an example we consider the traveling salesman problem: Given a set of cities and the distances between each pair of cities, the problem consists in finding the shortest route which starts from a city and visits each city exactly once and finally returns to the starting city. This problem is one of the so called "hard problems" in the sense that there exists no algorithm which solves the problem in polynomial time. The number of algorithms to solve this problem approximately is large and there exist a considerable literature on this topic. In the following we shall solve the problem for a small number of cities with "brute force".

Consider the following distance table of some cities in Switzerland:

|              | 1 | 2   | 3   | 4   | 5   | 6   | 7   | 8   | 9   |
|--------------|---|-----|-----|-----|-----|-----|-----|-----|-----|
| 1 Langenthal | 0 | 107 | 47  | 55  | 37  | 61  | 50  | 24  | 80  |
| 2 Brienz     |   | 0   | 77  | 53  | 117 | 147 | 115 | 109 | 83  |
| 3 Bern       |   |     | 0   | 112 | 85  | 97  | 42  | 73  | 136 |
| 4 Luzern     |   |     |     | 0   | 65  | 96  | 105 | 57  | 31  |
| 5 Aarau      |   |     |     |     | 0   | 53  | 77  | 14  | 71  |
| 6 Basel      |   |     |     |     |     | 0   | 90  | 46  | 108 |
| 7 Biel       |   |     |     |     |     |     | 0   | 66  | 129 |
| 8 Olten      |   |     |     |     |     |     |     | 0   | 78  |
| 9 Zug        |   |     |     |     |     |     |     |     | 0   |

The approximate distances (in km) were taken from Google Maps. We copy this distance table in a matrix and augment it to a symmetric Matrix

```
A =
     0    107     47     55     37     61     50     24     80
   107      0     77     53    117    147    115    109     83
    47     77      0    112     85     97     42     73    136
    55     53    112      0     65     96    105     57     31
    37    117     85     65      0     53     77     14     71
    61    147     97     96     53      0     90     46    108
```

| 50  | 115 | 42  | 105 | 77  | 90  | 0   | 66  | 129 |
|-----|-----|-----|-----|-----|-----|-----|-----|-----|
| 24  | 109 | 73  | 57  | 14  | 46  | 66  | 0   | 78  |
| 80  | 83  | 136 | 31  | 71  | 108 | 129 | 78  | 0   |

To solve the traveling salesman problem for this set of cities, we first choose a start city. Assume we start with number 1 (Langenthal). Then we have to generate all permutations of the cities 2–9 (Brienz–Zug) and for each permutations add up the distances from Langenthal and back again to Langenthal. One route would then be described for instance by the numbers

$$1 \to 4 \to 5 \to 8 \to 6 \to 9 \to 3 \to 2 \to 7 \to 1$$

and for the costs of this route (the sum of all km) we need to add the matrix elements

$$A(1,4) + A(4,5) + A(5,8) + A(8,6) + A(6,9) + A(9,3) + A(3,2) + A(2,7) + A(7,1)$$

which sum up to 666.

The function TravelSalesman is now not difficult to understand

```
function [solution,minimum]=TravelSalesman(A,startCity)
% Trvelsalesman solves the traveling salesman problem
%    The matrix A contains the distance table of cities. startCity is
%    the Number of city where the salesman starts and returns uses
%    Matlab's perms

[m,n]=size(A);
if m~=n,
   error('distance table is wrong')
end
if startCity>n
   error('Startcity not in the list')
end
c=[];                      % eliminate startCity from
for k=1:n                  % the list of cities
  if k~=startCity
    c=[c, k];
  end
end
T=perms(c);                % compute table T of all permutations
[m,p]=size(T)             % add startCity as first and last city
T=[ones(m,1)*startCity T ones(m,1)*startCity];
minimum=inf;
for k=1:m                  % for all permutations
  cost=0;
  for j=1:p+1              % compute the the cost
    cost=cost+A(T(k,j),T(k,j+1));
  end
  if cost<minimum         % save the minimum cost
    minimum=cost;
    solution=T(k,:);      % and the route
  end
end
```

**Fig. 9.4** Traveling salesman

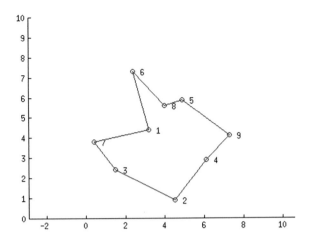

The following main program calls `TravelSalesman` and also plots the cities as points and the solution route, see Fig. 9.4.

```
% MainSales.m
% Main program for travelling salesman
A=[0 107 47  55  37  61  50 24  80  % distance matrix
   0 0    77  53 117 147 115 109  83
   0 0     0 112  85  97  42  73 136
   0 0     0   0  65  96 105  57  31
   0 0     0   0   0  53  77  14  71
   0 0     0   0   0   0  90  46 108
   0 0     0   0   0   0   0  66 129
   0 0     0   0   0   0   0   0  78
   0 0     0   0   0   0   0   0   0];
A=A+A';                                     % augment to sym. Matrix
clf
axis([0,10,0,10])
axis equal
hold
X=[3.2, 4.4                                 % cities as points on a map
   4.5, 0.9
   1.5, 2.4
   6.1, 2.9
   4.9, 5.9
   2.4, 7.3
   0.4, 3.8
   4.0, 5.6
   7.3, 4.1];
plot(X(:,1),X(:,2),'o')
n=9
for k=1:n
   text(X(k,1),X(k,2),['  ',num2str(k)])
end
[solution,minimum]=TravelSalesman(A,1)
plot(X(solution,1),X(solution,2))
```

```
>> MainSales
Current plot held
m =
      40320
p =
    8
solution =
    1    7    3    2    4    9    5    8    6    1
minimum =
   445
```

Starting with another city, e.g. with 2 (Brienz) yields the same result

```
>> [solution,minimum]=TravelSalesman(A,2)
m =
      40320
p =
    8
solution =
    2    4    9    5    8    6    1    7    3    2
minimum =
   445
```

which can be expected since there is exactly one optimal route.


## 9.3   Differential Equations

We shall consider in this section *ordinary differential equations* (ODEs). The solution of a *differential equation* is a *function*. Consider as example the equation

$$y'(x) = 2\ y(x).$$

We can guess that the solution is a exponential function since for this equation the derivative is a multiple of the function itself:

$$y(x) = e^{2x} \quad \Longrightarrow \quad y'(x) = 2\ e^{2x} = 2\ y(x).$$

But also $z(x) = ay(x)$ with some arbitrary constant $a$ is a solution. The equation has many solutions. To pick a specific solution we need to prescribe *initial conditions*. So if we consider the problem

$$y'(x) = 2\ y(x), \quad y(0) = 3$$

then the only solution is $y(x) = 3e^{2x}$.

   Differential equations have often solutions which cannot be represented by algebraic expressions. It is therefore necessary to consider numerical methods which compute approximations of the solutions.

A curve in the plane is best described in *parametric form*. For instance we describe an ellipse with semi-axes $a$ and $b$ by

$$x(t) = a\cos(t), \quad y(t) = b\sin(t), \quad 0 \le t \le 2\pi.$$

When we look for a curve in the plane, the differential equation is a *system of two equations* for the functions $x(t)$ and $y(t)$.

The following system of differential equations with the initial conditions $x(0) = 2$ and $y(0) = 0$ has as solution an ellipse

$$x'(t) = -2y(t)$$
$$y'(t) = \frac{x(t)}{2}.$$

We can verify this with the ansatz $x(t) = a\cos(t)$ and $y(t) = b\sin(t)$. It follows

$$x' = -a\sin t = -2b\sin t \implies a = 2b$$
$$y' = b\cos t = \frac{1}{2}a\cos t \implies b = \frac{1}{2}a.$$

Using the initial condition $x(0) = 2$ we get $a = 2$ and therefore $b = 1$. So the solution of the system is the ellipse

$$x(t) = 2\cos(t)$$
$$y(t) = \sin(t).$$

### 9.3.1  Numerical Integrator `ode45`

MATLAB provides many numerical integrators adapted for different types of ODEs (see `doc ode45`). A classic one if them is `ode45` an implementation of the explicit Runge–Kutta (4,5) pair of Dormand and Prince. It integrates the ODE with automatic step-size control that is it adapts the step-size such that the truncation error is kept constant.

In order to use a numerical integrator, the differential equation must be formulated in standard form as a *first order system of differential equations*

$$y' = f(t, y), \quad \text{with initial condition} \quad y(t_0) = y_0.$$

Example:

$$y''' + 5ty'' + y = e^{-t}, \quad y(0) = 10, \; y'(0) = 0, \; y''(0) = -0.1.$$

This third order differential equation is transformed to a first order system by intro-
ducing new variables $z_1(t) = y(t)$, $z_2(t) = y'(t)$ and $z_3(t) = y''(t)$. Then by differ-
entiating and replacing the $y$ we get the system

$$
\begin{aligned}
z_1' &= y' = z_2 \\
z_2' &= y'' = z_3 \\
z_3' &= y''' = -5ty'' - y + e^{-t} = -5tz_3 - z_1 + e^{-t}
\end{aligned}
$$

which written in matrix-vector notation is

$$
z' = Az + b \quad A = \begin{pmatrix} 0 & 1 & 0 \\ 0 & 0 & 1 \\ -1 & 0 & -5t \end{pmatrix}, b = \begin{pmatrix} 0 \\ 0 \\ e^{-t} \end{pmatrix}.
$$

The initial conditions for this system are $z(0) = [10, 0, -0.1]^\top$.

The MATLAB function ode45 can be used to solve such a first order system of
ODEs. We need to define the system as a function odefun. For a scalar t and a
vector $y$, odefun(t,y) must return a column vector corresponding to $f(t, y)$.
Then the ODE is integrated with

```
[tout,yout] = ode45(odefun,tspan,y0)
```

where tspan indicates the interval for which the functions should be computed, so
for instance we could have tspan=[0,10]. The third parameter y0 contains the
values of the initial conditions. The output parameters are [tout,yout]. Each
row in the solution array yout corresponds to the function values $z_1, \ldots, z_n$ at a
time returned in the column vector tout.

For our example we first program the function

```
function dz=fsystem(t,z)
A=[0 1 0
   0 0 1
   -1 0 -5*t];
b=[0 0 exp(-t)]';
dz=A*z+b;
```

The main program is then

```
% Example for ode45
t=0; y0=[10 0 -0.1]';
[tt,yy]=ode45(@fsystem,[0,5],y0)
plot(tt,yy)
```

It produces a table of results and a plot of the functions $z_1(t) = y(t)$ (blue color),
$z_2(t) = y'(t)$ (green color) and $z_3(t) = y''(t)$ (red color), see Fig. 9.5.

**Fig. 9.5** Third order ODE

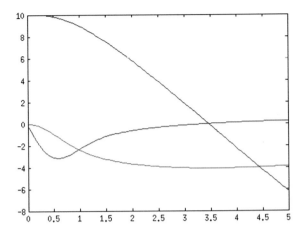

## 9.3.2   Dog Attacking a Jogger

This example is taken from [4]. We consider the following problem: while a jogger is running on some trail in the plane, he is being attacked by a dog. Compute the orbit $x(t)$, $y(t)$ of the dog.

We assume that the dog is running full speed with constant velocity $w$. His velocity vector points at every time to its goal, the jogger. We assume that the motion of the jogger is described by the two functions $X(t)$ and $Y(t)$. The following equations hold:

1. $\dot{x}^2 + \dot{y}^2 = w^2$: The dog is running with constant speed.
2. The velocity vector of the dog is parallel to the difference vector between the position of the jogger and the dog:

$$\begin{pmatrix} \dot{x} \\ \dot{y} \end{pmatrix} = \lambda \begin{pmatrix} X - x \\ Y - y \end{pmatrix} \quad \text{with } \lambda > 0.$$

If we substitute this in the first equation we obtain

$$w^2 = \dot{x}^2 + \dot{y}^2 = \lambda^2 \left\| \begin{pmatrix} X - x \\ Y - y \end{pmatrix} \right\|^2.$$

This equation can be solved for $\lambda$:

$$\lambda = \frac{w}{\left\| \begin{pmatrix} X-x \\ Y-y \end{pmatrix} \right\|} > 0.$$

Finally, substitution of this expression for $\lambda$ in the second equation yields the differential equation of the orbit of the dog:

$$\begin{pmatrix} \dot{x} \\ \dot{y} \end{pmatrix} = \frac{w}{\left\| \begin{pmatrix} X-x \\ Y-y \end{pmatrix} \right\|} \begin{pmatrix} X-x \\ Y-y \end{pmatrix}. \tag{9.1}$$

To solve this system we need to program several functions. First the trail of the jogger. We let him run on the $x$-axis:

```
function s=jogger1(t);
s=[8*t; 0];
```

Next we program the ODE for the dog:

```
function dz=dog1(t,z)
global w
X=jogger1(t);
h=X-z;
nh=norm(h);
dz=w/nh*h;
```

The main program then becomes

```
% mainDog1
global w
y0=[60;70];                     % starting point of the dog
w=10;                           % w   speed of the dog
[t,Y]=ode45(@dog1,[0,10],y0)
clf; hold on;
axis([-10,100,-10,70]);
plot(Y(:,1),Y(:,2));
J=[];                           % recompute the trail of the
for k=1:length(t),             % jogger for the same points
  w=jogger1(t(k));              % in time as used for the dog
  J=[J; w'];
end;
plot(J(:,1), J(:,2),'r');
```

To plot the trail of the jogger we recompute it using the same points in time as were used for the dog's orbit. This will be also useful for showing the movements (see below). We get the following result, see Fig. 9.6. Since the dog is running faster (with speed 10) than the jogger (with speed 8), by integrating a little longer the dog should catch the jogger. However, the system (9.1) has a singularity when the dog reaches the jogger: the norm of the difference vector becomes zero and we should stop the integration, since also the numerical integrator gets in trouble.

We do not know the exact time when this happens, so we should stop integrating when the dog is near the jogger. For such situations MATLAB provides the possibility to define another termination criterion for the integration, different from a given upper bound for the independent variable. It is possible to terminate the integration by checking zero crossings of a function.

**Fig. 9.6** Dog orbit (*blue*), jogger trail (*red*)

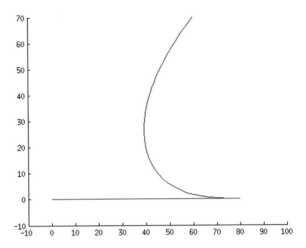

To do so we need to add in the main programm a call to the function `odesets`

```
options= odeset('events','on')
```

to trigger MATLAB to observe events. The paramenter `options` has to be passed to `ode45` as additional input parameter:

```
[t,Y]=ode45('dog2',[0,30],y0,options)
```

Furthermore the function describing the dog's movement has to be adapted, we would like to terminate integration when $||(X - x, Y - y)^\top||$ becomes small. In order to do so we have to add a third input and two more output parameters to the function describing the dog.

```
function [dz,isterminal,direction]=dog2(t,z,flag);
%
global w                    % w = speed of the dog
X=jogger1(t);
h=X-z;
nh=norm(h);
if nargin<3|isempty(flag)  % normal output
  dz=(w/nh)*h;
else
  switch(flag)
    case 'events'          % at norm(h)=0 there is a singularity
      dz= nh-1e-3;         % zero crossing at pos_dog=pos_jogger
      isterminal= 1;       % this is a stopping event
      direction= 0;        % don't care if decrease or increase
    otherwise
      error(['Unknown flag: ' flag]);
  end
end
```

The integrator `ode45` calls the function in two ways: The first one consists of dropping the third parameter. The function then returns only the parameter `dz`: the speed of the dog. In the second way the keyword `'events'` is assigned to the parameter `flag`. This keyword tells the function to return the zero-crossing function in the first output `dz`. The second output `isterminal` is a logical vector that tells the integrator, which components of the first output force the procedure to stop when they become zero. Every component with this property is marked with a nonzero entry in `isterminal`. The third output parameter `direction` is also a vector that indicates for each component of `dz` if zero crossings shall only be regarded for increasing values (`direction` = 1), decreasing values (`direction` = -1) or in both cases (`direction` = 0). The condition for zero crossings is checked in the integrator. The main program becomes

```
% mainDog2
global w
y0=[60;70];                    % starting point of the dog
w=10;                          % w   speed of the dog
options=odeset('Events','on')
[t,Y]=ode45('dog2',[0,10],y0,options)
clf; hold on;
axis([-10,120,-10,80]);
plot(Y(:,1),Y(:,2));
J=[];
for k=1:length(t),
  w=jogger1(t(k));
  J=[J; w'];
end;
plot(J(:,1), J(:,2),'*r');
```

and we get the Fig. 9.7.

I would be nice to show the movement dynamically. For this we plot for each time interval piecewise the orbit of the dog and the trail of the jogger. In order to not delay the plotting we have to include the command `drawnow` in the loop. We also include a `pause`-statement to slow down the computation, so that the movements become nicely visible. Thus the program becomes

**Fig. 9.7** Integrating till dog reaches the jogger

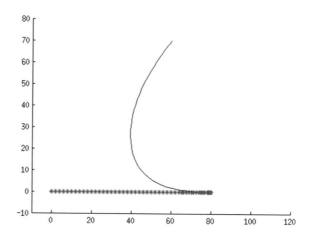

```
% mainDog3
global w
y0=[60;70];                      % starting point of the dog
w-10;                            % w  speed of the dog
options=odeset('Events','on')
[t,Y]=ode45('dog2',[0,10],y0,options)
clf; hold on;
axis([-10,120,-10,80]);
J=[];
for k=1:length(t),
  w=jogger1(t(k));
  J=[J; w'];
end;
title ('Dog Attacking Jogger');
for h=1:length(t)-1,
  plot ([Y(h,1),Y(h+1,1)],[Y(h,2),Y(h+1,2)]);
  plot ([J(h,1),J(h+1,1)],[J(h,2),J(h+1,2)],'r')
  drawnow;
  pause(0.05);
end
hold off;
```

## 9.4  MATLAB-Elements Used in This Chapter

**sum:**      Sum of array elements

If A is a vector, then sum(A) returns the sum of the elements.
If A is a matrix, then sum(A) returns a row vector containing the sum of each column.

**perms:**    All possible permutation

P = perms(v) returns a matrix containing all permutations of the elements of vector v in reverse lexicographic order. Each row of P contains a different permutation of the n elements in v. Matrix P has the same data type as v, and it has n! rows and n columns.

**ode45:**    Solve nonstiff differential equations; medium order method

[T,Y] = ode45(odefun,tspan,y0)
odefun: A function handle that evaluates the right side of the differential equations y' = f(t,y)
tspan: A vector specifying the interval of integration, [t0,tf]. ode45 imposes the initial conditions at tspan(1), and integrates from tspan(1) to tspan(end). To obtain solutions at specific times (all increasing or all decreasing), use tspan = [t0,t1,…,tf].
y0: A vector of initial conditions.
[T,Y] = solver(odefun,tspan,y0,options)

options: Structure of optional parameters that change the default integration properties. This is the fourth input argument.

You can create options using the odeset function. See odeset for details.

**odeset:** Create or alter options structure for ordinary differential equation solvers

options = odeset('name1',value1,'name2',value2,...) creates an options structure that you can pass as an argument to the ode45 solver. In the resulting structure, options, the named properties have the specified values. For example, 'name1' has the value value1. Any unspecified properties have default values. It is sufficient to type only the leading characters that uniquely identify a property name. Case is ignored for property names.

## 9.5  Problems

1. Waiting for the elevator. We consider a building with $n$ floors. A elevator is serving these floors and we are interested to know the distribution of the waiting time from pressing the elevator button till the elevator opens the door.

    Assume the time unit for the moving of the elevator one floor is one. We make $m$ experiments in which the elevator is randomly located on one floor and the person is also coming randomly on one floor. The difference of the two floors is proportional to the time the elevator needs to come.

    Perform $m = 10'000$ experiments for a $n = 50$ floors building. Construct and plot the histogram of the waiting times.

2. Given a set of points in the unit square. Write a program which computes and plots the two closest points.

    (a) Write a function [P,Q,minimum]=ClosestPoints(x,y) which computes all the distances between two points and stores the minimal distance and the two points $P$ and $Q$ which are closest.
    (b) Generate $n$ points $(x_k, y_k)$ using the function rand. Then call the function ClosestPoints, plot the points and mark the two closest points by coloring them differently.

3. Shortest distance between two point sets:

    (a) Consider the circle with center $(5, 6)$ and radius $r = 2$ and the ellipse with center at origin and $a = 1$ and $b = 0.5$ parallel to the coordinate axis.
    Sample points on the ellipse and on the circle. Compute by brute force a point $P$ on the circle and a point $Q$ on the ellipse with minimal distance.
    (b) The circle with center $(5, 6)$ and radius $r = 2$ and the ellipse with center $(4, 4)$, $a = 2$ and $b = 3$ intersect. Try to compute the intersection points by brute force.

4. Knapsack Problem: given a bag with a given maximum load limit $W$. Put in that bag items from the following table in order to maximize the sum of the value of the items but not exceeding the total weight $W$:

| item | 1 | 2 | 3 | 4 | 5 | 6 | 7 |
|---|---|---|---|---|---|---|---|
| weights | 3.3 | 4.6 | 1.7 | 5.8 | 7.7 | 3.1 | 5.3 |
| values | 7 | 9 | 5 | 12 | 14 | 6 | 12 |

Write a brute force program that solves the problem for a collections of bags:

$$W = [8, 10, 11, 15, 20, 21, 25, 26, 30, 32]$$

5. A dog would like to cross a river of width $b$. He starts at point $(b, 0)$ with the goal to swim to $(0, 0)$ where he has detected a sausage. His swim velocity $v_D$ is constant and his nose points always to the sausage. The river flows north in direction of the $y$-axis and velocity of the flow of the river $v_R$ is everywhere constant.

(a) Develop the differential equation describing the orbit $z(t) = (x(t), y(t))^\top$ of the dog.
(b) Program a MATLAB function zp=dog(t,z) which describes the differential equation. The velocities $v_D$ and $v_R$ may be declared as global variables.
(c) Use the program quiver and plot the slope field for $b = 1$, $v_R = 1$ and the following three cases for the dog velocity $v_D = 0.8$, 1.0 and 1.5.
Note: quiver(X,Y,Xp,Yp) needs 4 matrices. X and Y contain the coordinates of the points and Xp and Yp the two components of the velocity at that point. To compute these you can use the function dog e.g.

```
z=dog(0,[X(k,j),Y(k,j)]);Xp(k,j)=z(1);Yp(k,j)=z(2);
```

(d) Develop a MATLAB integrator for the method of Heun of order 2

```
function Z= OdeHeun(f,z0,tend,n)
% ODEHEUN integrates y'=f(t,y), y(0)=z0 with Heun
%    from  t=0 to tend using a fixed step size h=tend/n
```

which integrates a given system of differential equations $y' = f(t, y)$ and stores the results in the matrix $Z$. The $i$ th row of the matrix $Z$ contains the values

$$[t_i, y_1(t_i), \ldots, y_n(t_i)].$$

Compute and plot the orbits for the three dog velocities. You may want to stop the integration before executing all $n$ steps when the dog arrives close to the origin or in the case when $v_D < v_R$ the dog is near the $y$-axis.

**Hint:** An approximation $y_k \approx y(t_k)$ of the solution of the differential equation $y' = f(t, y)$, $y(t_0) = y_0$ is computed for constant stepsize $h$ by the method of Heun with the following statements

$$k = 0, 1, 2, \ldots$$
$$k_1 = f(t_k, y_k)$$
$$y^* = y_k + hk_1$$
$$k_2 = f(t_k + h, y^*)$$
$$t_{k+1} = t_k + h$$
$$y_{k+1} = y_k + \frac{h}{2}(k_1 + k_2)$$

# Chapter 10
# Solutions of Problems

## 10.1  Chapter 1: Starting

1. Start MATLAB with the GUI and watch the introductory video and study the tutorial.
2. If you own a computer or laptop without MATLAB then download and install the open source software GNU OCTAVE on it.

## 10.2  Chapter 2: How a Computer Calculates

1. Consider the following finite decimal arithmetic: 2 digits for the mantissa and one digit for the exponent. So the machine numbers have the form $\pm Z.ZE\pm Z$ where $Z \in \{0, 1, \ldots, 9\}$

   (a) How many normalized machine numbers are available?
   (b) Which is the overflow- and the underflow range?
   (c) What is the machine precision?
   (d) What is the smallest and the largest distance of two consecutive machine numbers?

   Solution:

   - We first count the machine numbers. We can form 19 different exponents: $-9, -8, \ldots, 9$. The first digit, before the decimal point, can not be zero, because we consider only normalized numbers, thus we have 9 possibilities for the first digit. Thus in total we have $2 \times 9 \times 10 \times 19 = 3420$ normalized machine numbers plus the number zero. Therefore the grand total is 3421 machine numbers.
   - The largest number is $9.9E9 = 9,900,000,000$ and the smallest positive number is $1.0E-9$. The overflow range is $|x| > 9.9E9$ and the underflow range is $0 < |x| < 1.0E-9$.

© Springer International Publishing Switzerland 2015                                   99
W. Gander, *Learning MATLAB*, UNITEXT - La Matematica per il 3+2 95,
DOI 10.1007/978-3-319-25327-5_10

- The machine precision is the spacing between the numbers in $(1, 2)$ thus $\varepsilon = 1.1\text{E}0 - 1.0\text{E}0 = 1\text{E}{-}1$.
- The largest distance between two machine numbers occurs when the exponent is 9: $9.9\text{E}9 - 9.8\text{E}9 = 1\text{E}8$. The smallest distance is $1.1\text{E}{-}9 - 1.0\text{E}{-}9 = 1\text{E}{-}10$.

2. Solving a quadratic equation: Write a MATLAB function

```
function [x1,x2]=QuadraticEq(p,q)
```

which computes the real solutions of an equation

$$x^2 + px + q = 0.$$

If the solutions turn out to be complex then write an error message. Test your program with the following examples:

- $(x - 2)(x + 3) = x^2 + x - 6 = 0$   thus $p = 1$ and $q = -6$.
- $(x - 10^9)(x + 2 \cdot 10^{-9}) = x^2 + (2 \cdot 10^{-9} - 10^9)x + 2$
  thus $p = 2e{-}9 - 1e9$ and $q = -1e9$.
- $(x + 10^{200})(x - 1) = x^2 + (10^{200} - 1)x - 10^{200}$
  thus $p = 1e200 - 1$ and $q = -1e200$.

Comment your results.

Solution: Using the textbook formula

$$x_{1,2} = -\frac{p}{2} \pm \sqrt{\left(\frac{p}{2}\right)^2 - q}$$

we obtain the function

```
function [x1,x2]=QuadEquationNaive(p,q)
discriminant=(p/2)^2-q;
if discriminant<0
    error('solutions are complex')
end
d=sqrt(discriminant);
x1=-p/2+d; x2=-p/2-d;
```

We test this function:

- $(x - 2)(x + 3) = x^2 + x - 6 = 0$

  ```
  >> [x1,x2]=QuadEquationNaive(1,-6)
  ```

  ```
  x1=2, x2=-3      correct
  ```
- $(x - 10^9)(x + 2 \cdot 10^{-9}) = x^2 + (2 \cdot 10^{-9} - 10^9)x - 2$

  ```
  >> [x1,x2]=QuadEquationNaive(2e-9-1e9,-2)
  ```

  ```
  x1=1.0000e+09, x2=0      wrong
  ```

- $(x + 10^{200})(x - 1) = x^2 + (10^{200} - 1)x - 10^{200}$

```
>> [x1,x2]=QuadEquationNaive(1e200-1,-1e200)

x1=Inf, x2=-Inf    wrong
```

Why do we get wrong answers? When looking at the textbook formula we notice that for large $|p|$ forming $p^2$ may overflow. This is the case in the third example. On the other hand for small $q$ the formula is

$$x_{1,2} = -\frac{p}{2} \pm \sqrt{\left(\frac{p}{2}\right)^2 - q} \approx -\frac{p}{2} \pm \frac{p}{2}$$

and one solution is affected by cancellation. This is the case in the second example.

We can avoid the overflow by factoring out. The cancellation can be avoided by computing first the solution which has the larger absolute value and then use the relation of Vieta:

$$x_1 x_2 = q$$

to compute the smaller solution without cancellation. Thus instead of the textbook formula we use

$$x_1 = -\text{sign}(p)\left(|p|/2 + |p|\sqrt{\frac{1}{4} - q/p/p}\right)$$

$$x_2 = q/x_1 \qquad \text{Vieta}$$

We obtain the function

```
function [x1,x2]=QuadraticEq(p,q)
if abs(p/2)>1                            % avoid overflow
  factor=abs(p); discriminant=0.25-q/p/p; % by factoring out
else
  factor=1; discriminant=(p/2)^2-q;
end
if discriminant<0
  error('Solutions are complex')
else
  x1=abs(p/2)+factor*sqrt(discriminant);  % compute larger solution
  if p>0, x1=-x1; end                      % adapt sign
  if x1== 0, x(2)=0;
  else
    x2=q/x1;                              % avoid cancellation
  end                                      % for smaller solution
end
```

This time we get

- $(x - 2)(x + 3) = x^2 + x - 6 = 0$

```
>> [x1,x2]=QuadraticEq(1,-6)    x1 = 2    x2 = -3
```

correct

- $(x - 10^9)(x + 2 \cdot 10^{-9}) = x^2 + (2 \cdot 10^{-9} - 10^9)x - 2$

```
>> [x1,x2]=QuadraticEq(2e-9-1e9, 2)   x1-1.0000e+09   x2=-2.0000e-09
```

correct!

- $(x + 10^{200})(x - 1) = x^2 + (10^{200} - 1)x - 10^{200}$

```
>> [x1,x2]=QuadraticEq(1e200-1,-1e200)     x1=-1.0000e+200    x2=1
```

correct!

## 10.3   Chapter 3: Plotting Functions and Curves

1. We are given the points

$$
\begin{array}{c|cccccc}
x & 0.9 & 2.3 & 3.9 & 4.6 & 5.8 & 7.3 \\
\hline
y & 2.9 & 4.1 & 4.8 & 7.0 & 7.0 & 8.7
\end{array}
$$

(a) Define a region to plot the points using `axis`. Use `hold` to freeze the axis.
(b) Plot the points using the symbol 'x'.
(c) We want to fit a regression line through the points, that means compute the parameters $a$ and $b$ such that

$$ y_k = ax_k + b, \quad k = 1, \ldots, 6. $$

This is a linear system of equations with two unknowns and 6 equations. It cannot be solved exactly, the equations contradict themselves. However, the MATLAB \-operator does solve the system in the least squares sense by computing the best approximation for all equations.
Form the linear system $A\binom{a}{b} = y$ and solve it by A\y
(d) Using the computed values of $a$ and $b$, plot the regression line on the same plot with the points.

Solution:

```
x=[0.9 2.3 3.9 4.6 5.8 7.3]'
y=[2.9 4.1 4.8 7.0 7.0 8.7]'

axis([0,8,0,9])
hold
plot(x,y,'x')
A=[x,ones(size(x))]
z=A\y
a=z(1); b=z(2);
plot(x,a*x+b)
```

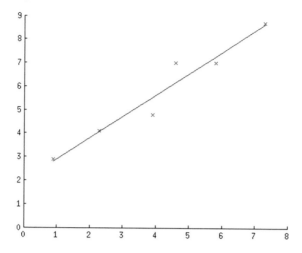

2. Ellipse plots.

   (a) Plot the ellipse with center in origin and the main axis $a = 3$ on the $x$-axis
       and minor axis $b = 1$. Plot also the center using the symbol '+'.
   (b) Now move the ellipse so that the center is the point $(4, -1)$ and the direction
       of the main axis has an angle of $-30°$ with the $x$-axis. Plot this new ellipse
       in the same frame.

   **Hint**: Use a rotation matrix of the form

   $$Q = \begin{pmatrix} \cos\alpha & -\sin\alpha \\ \sin\alpha & \cos\alpha \end{pmatrix}$$

   to rotate the coordinates of the ellipse.

   Solution:

```
clear,clf
axis([-4,8,-4,8])
axis equal                   % both axes same units
hold
a=3, b=1
t=0:0.01:2*pi;
x=a*cos(t)
y=b*sin(t)
pause
plot(0,0,'+')                % plot first ellipse with center
plot(x,y)
alpha=-pi/6    % -30 degrees   % rotation angle
c=cos(alpha), s=sin(alpha)
Q=[c -s; s c]                % rotation matrix
pause
E=ones(2,length(x));
```

```
z=diag([4,-1])*E+Q*[x;y]
plot(z(1,:),z(2,:))                 % plot second ellipse
plot(4,-1,'x')
```

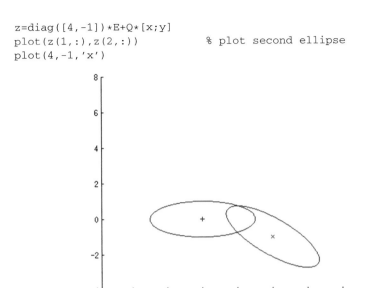

3. Plot for $-3 \le x \le 3$ and $-5 \le y \le 5$ the function $f(x, y) = x^2 - 2yx^3$ using
   `contour` and `mesh`.

Solution:

```
clear, clf
[x,y]=meshgrid(-3:0.1:3,-5:0.1:5);
z=x.^2-2*y.*x.^3;
figure(1)
mesh(x,y,z)
figure(2)
contour(x,y,z,250)
```

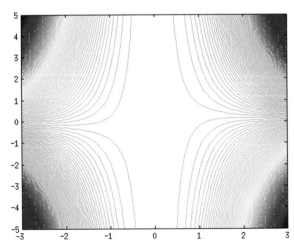

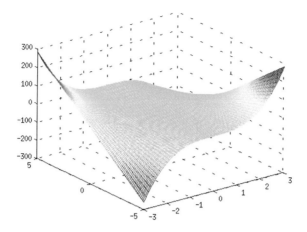

## 10.4   Chapter 4: Some Elementary Functions

1. Explain what happens in Algorithm e1 when $x = -20$.
   **Hint**: look at the size of the largest term and at the final result. What happens when computing the result in finite arithmetic?

   Solution: For large negative $x$, e.g. for $x = -20$ and $x = -50$, we obtain using the function e1

```
>> e1(-20)
ans =      5.621884807271559e-09
>> exp(-20)
ans =      2.061153622438558e-09
>> e1(-50)
ans =      1.107293448191918e+04
>> exp(-50)
ans =      1.928749847963918e-22
```

   which are completely incorrect. The reason is that for $x = -20$, the terms in the series

$$1 - \frac{20}{1!} + \frac{20^2}{2!} - \cdots + \frac{20^{20}}{20!} - \frac{20^{21}}{21!} + \cdots$$

   become large and have alternating signs. The largest terms are

$$\frac{20^{19}}{19!} = \frac{20^{20}}{20!} = 4.3e7.$$

   The partial sums should converge to $e^{-20} = 2.06e-9$. But because of the growth of the terms, the partial sums become large as well and oscillate as shown in

**Fig. 10.1** Partial sums of the Taylor expansion of $e^{-20}$

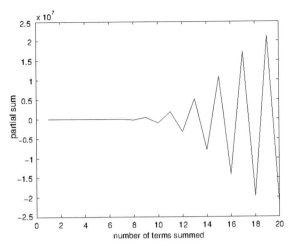

**Table 10.1** Numerically computed partial sums of $e^{-20}$

| Number of terms summed | Partial sum |
|---|---|
| 20 | $-2.182259377927747e+07$ |
| 40 | $-9.033771892137873e+03$ |
| 60 | $-1.042344520180466e-04$ |
| 80 | $6.138258384586164e-09$ |
| 100 | $6.138259738609464e-09$ |
| 120 | $6.138259738609464e-09$ |
| Exact value | $2.061153622438558e-09$ |

Fig. 10.1. Table 10.1 shows that the largest partial sum has about the same size as the largest term. Since the large partial sums have to be *diminished by additions/subtractions of terms*, this cannot happen without cancellation. Neither does it help to first sum up all positive and negative parts separately, because when the two sums are subtracted at the end, the result would again suffer from catastrophic cancellation. Indeed, since the result

$$e^{-20} \approx 10^{-17} \frac{20^{20}}{20!}$$

is about 17 orders of magnitude smaller than the largest intermediate partial sum and the IEEE Standard has only about 16 decimal digits of accuracy, we cannot expect to obtain even one correct digit! To obtain a correct value to 16 digits we would have to compute with over 30 digits.

2. Write a MATLAB-function to compute $\sin x$ using the series (4.7). In order to avoid cancellation for large $|x|$ reduce the argument to the interval $[-\pi/2, \pi/2]$.

Solution:

The reduction of the arguments $x$ to the interval $[-\pi/2, \pi/2]$ is important because the series is alternating and the summation will be affected by catastrophic cancellation for large $|x|$. We first reduce the angle to $[0, 2\pi]$ using the mod-function and then in a second step to $[-\pi/2, \pi/2]$. For very large arguments the reduction by the mod-function will be inaccurate. So we must expect then a less accurate result.

```
function s=MySin(x);
% MYSIN Machine-independent computation of sin(x)
% using Taylor Series
if x<0, v=-1; x=-x; else v=1;end   % store sign
x=mod(x,2*pi);                     % reduce angle to [0,2*pi]
if x>pi, x =x-pi; v=-v; end        % further reductions
if x>pi/2, x=pi-x; end             % so that
if v==-1, x=-x; end                % -pi/2<x<pi/2
s=x; t=x; i=1; sold=2*x;           % sum the series
while s~=sold,
   sold=s; i=i+2;
   t=-t*x/(i-1)*x/i;
   s=sold+t;
end

>>  x=1000; [MySin(x)-sin(x)]
ans =
   2.1760e-14
>>  x=10; [MySin(x)-sin(x)]
ans =
  -2.2204e-16
>>  x=1e10; [MySin(x)-sin(x)]
ans =
   3.4036e-07
>>  x=-1000; [MySin(x)-sin(x)]
ans =
  -2.1760e-14
```

3. Do the same for $\cos x$.

Solution: We proceed similarly as with MySin. The reduction needs adjustment.

```
function s=MyCos(x)
% MYCOS Machine-independent computation of cos(x)
% using Taylor Series
x=abs(x); v=1;                     % cos is symmetric
x=mod(x,2*pi);                     % reduce angle to [0,2*pi]
if x>3*pi/2,
   x =x-2*pi;                      % further reductions
```

```
elseif x>pi/2,                        % so that
   x=x-pi; v=-1;                       % -pi/2<x<pi/2
end
s=1; t=1; i=0; sold=2;                 % sum the series
while s~=sold,
   sold=s; i=i+2;
   t=-t*x/(i-1)*x/i;
   s=sold+t;
end
if v==-1,s=-s;end                      % adjust sign of function

>> x=1; [MyCos(x)-cos(x)]
ans =
   -1.1102e-16
>> x=10; [MyCos(x)-cos(x)]
ans =
   2.2204e-16
>> x=-100; [MyCos(x)-cos(x)]
ans =
   2.1094e-15
>> x=1000; [MyCos(x)-cos(x)]
ans =
   -3.2196e-14
>> x=1e10; [MyCos(x)-cos(x)]
ans =
   1.9004e-07
```

4. Combine both functions and write a function to compute $\tan x$.

   Solution: we just use the relation $\tan x = \frac{\sin x}{\cos x}$

```
function y=MyTan(x)
% MYTAN    compute tan(x) using only the four basic operations.
%   refers to MySin and MyCos.
y=MySin(x)/MyCos(x);

>> x=1e10; [MyTan(x)-tan(x)]
ans =
   5.1134e-07
>> x=300; [MyTan(x)-tan(x)]
ans =
   2.3782e-11
>> x=30; [MyTan(x)-tan(x)]
ans =
   5.0626e-14
```

5. Write a function to compute $\arctan x$ for $|x| < 1$ using the series (4.8) and compare your result with the standard MATLAB-function atan(x).

   Solution:

```
function s=MyArctan(x)
% MYARCTAN computes the function arctan(x)
```

```
%  for |x|<1
s=x; t=x; k=1; sold=0;
while s~=sold
   sold=s; k=k+2; t=-t*x^2;
   s=s+t/k;
end

>> for x=[0.7,-0.7,0.5,-0.5,0.1,-0.1]
      [MyArctan(x)-atan(x)]
   end
ans =
   1.1102e-16
ans =
  -1.1102e-16
ans =
  -2.2204e-16
ans =
   2.2204e-16
ans =
  -2.7756e-17
ans =
   2.7756e-17
```

## 10.5  Chapter 5: Computing with Multiple Precision

For the following problems, make use of the functions we developed for computing
Euler's number $e$.

1. Compute using multiple precision the powers of 2:

$$2^i, \quad i = 1, 2, \ldots, 300$$

Solution:

```
function Power2(m)
% POWER2 computes the powers of 2 in multiple precision
%   2^k, k=0, .., m
n=round(m*log10(2))+1;       % compute how many array elements we need
c=10;                        % we pack one digit in one element
a=zeros(1,n,'uint32');
a(n)=1;
for k=1:m
   a=a*2;                    % generate next power of 2
   a=Carry(c,a);
   ['2^',sprintf('%01d',k),  ' = ', sprintf('%01d',a)]
end
```

2. Write a program to compute factorials using multiple precision:

$$n!, \quad n = 1, 2, \ldots, 200$$

Solution: It is not simple to predict how many decimal digits are needed to represent the number $n$! The Scottish mathematician James Stirling derived the asymptotic formula

$$n! \sim \sqrt{2\pi n}\left(\frac{n}{e}\right)^n.$$

We can estimate using `log10` the number of digits. Don't just take the logarithm of the above expression! It will become already `Inf` for $n < 200$. We have to split the expression in several logarithms

$$\log 10(n!) = \log 10(\sqrt{2\pi n}) + n \log 10(n) - n \log 10(e)$$

to avoid overflow.

```
function Factorials(n)
% FACTORIALS computes k! for k=1,...,n in multiple precision
% use Stirlings formula to estimate number of digits
m=round(log10(sqrt(2*pi*n))+n*(log10(n)-log10(exp(1))))
c=10;                     % we pack one digit in one element
a=zeros(1,m,'uint32');
a(m)=1;
for k=1:n
   a=a*k;                 % generate k!
   a=Carry(c,a);
   [sprintf('%01d',k),   '!  = ', sprintf('%01d',a)]
end
```

3. Compute $\pi$ to 1000 decimal digits. Use the relation by C. Størmer:

$$\pi = 24 \arctan\frac{1}{8} + 8 \arctan\frac{1}{57} + 4 \arctan\frac{1}{239}.$$

**Hints**:

- Compute first a multiprecision arctan function using the Taylor-series (4.8) as proposed in Chap. 4:

$$\arctan x = \sum_{k=0}^{\infty}(-1)^k\frac{x^{2k+1}}{2k+1} = x - \frac{x^3}{3} + \frac{x^5}{5} - \cdots$$

- The above series is alternating so there is a danger of cancellation. However, since it is used only for $|x| < 1$ this is not much a concern. What we need is a new function `Sub`

```
function r=Sub(c,a,b)
% SUB computes r=a-b where a and b are multiprecision numbers
% with a>b.
```

to subtract two multiprecision numbers. One has to be careful not to generate negative numbers, all intermediate results have to remain positive.
- To compute $\pi$ we have to evaluate for some integer $p > 1$ the function $\arctan(1/p)$. When generating the next term after

$$t_k = \frac{x^{2k+1}}{2k+1}$$

for $x = 1/p$ we have to form

$$t_{k+1} = t_k/p^2/(2k+1).$$

There is bug that one has to avoid: by dividing the last term twice by $p$ and a third time by $2k + 1$ the variable imin is updated. For the next term we need to know the value of imin before the division by $2k + 1$! Otherwise we will get erroneous results when forming $t_k/p^2$.

Solution:

For the subtraction of two multiple precision numbers we propose

```
function r=Sub(c,a,b)
% SUB computes r=a-b where a and b are multiprecision numbers
% with a>b.
n=length(a);
r=a;
for i=n:-1:1
  while a(i)<b(i)                        % need  to borrow from left
    a(i)=a(i)+c; b(i-1)=b(i-1)+1;
  end
  r(i)=a(i)-b(i);
end
```

We reuse the functions Divide, Add and Carry. The multiprecision function for arctan becomes

```
function s=AtanMultPrec(c,n,p)
% ATANMULTPREC computes n*log10(c) decimal digits
%   of the function value s=arctan(1/p) where p>1 is
%   an integer number
s=zeros(1,n,'uint32');
s(1)=1;
imin=0;                              % imin counts leading zeros in t
[s,imin]=Divide(c,imin,p,s);         % s=1/p
t=s;                                 % first term
k=1;
sig=1;                               % the sign of the term
while imin<n
  k=k+2;
  [t,imin]=Divide(c,imin,p^2,t);% new nominator of term
```

```
  h=Divide(c,imin,k,t);              % division without change of imin
  sig=-sig;                          % change sign
  switch sig
    case -1
      s=Sub(c,s,h);                  % subtract or
    case 1
      s=Add(imin,s,h);               % add term h to s
      s=Carry(c,s);
  end
end
```

Finally we compute $\pi$ by the function

```
function s=Pi(c,n)
% PI computes n decimal digits of pi using the
%   formula of Stormer.
s=24*AtanMultPrec(c,n,8);       % generate the 3 terms
t2=8*AtanMultPrec(c,n,57);
t3=4*AtanMultPrec(c,n,239);
s=Add(1,s,t2);                  % add the terms
s=Add(1,s,t3);
s=Carry(c,s);
sprintf('%01d',s)
```

and get for $n = 1000$

```
>> tic, Pi(10,1000); toc
ans =
3141592653589793238462643383279502884197169399
3751058209749445923078164062862089986280348253
4211706798214808651328230664709384460955058223
1725359408128481117450284102701938521105559644
6229489549303819644288109756659334461284756482
3378678316527120190914564856692346034861045432
6648213393607260249141273724587006606315588174
8815209209628292540917153643678925903600113305
3054882046652138414695194151160943305727036575
9591953092186117381932611793105118548074462379
9627495673518857527248912279381830119491298336
7336244065664308602139494639522473719070217986
0943702770539217176293176752384674818467669405
1320005681271452635608277857713427577896091736
3717872146844090122495343014654958537105079227
9689258923542019956112129021960864034418159813
6297747713099605187072113499999983729780499510
5973173281609631859502445945534690830264252230
8253344685035261931188171010003137838752886587
5332083814206171776691473035982534904287554687
3115956286388235378759375195778185778053217122
6806613001927876611195909216419964
```

```
Elapsed time is 35.828138 seconds.
```

## 10.6   Chapter 6: Solving Linear Equations

1. LU-decomposition Consider the linear system $Ax = b$ defined by the matrix

```
>> format short e, format compact
>> n=5; A=invhilb(n), b=eye(n,1)
```

(a) Apply Gaussian Elimination (without pivoting) to reduce the system to
$Ux = y$

```
for j=1:n-1                          % Elimination
    for k=j+1:n
      fak=A(k,j)/A(j,j);
      A(k,j:n)=A(k,j:n)-fak*A(j,j:n);
      b(k)=b(k)-fak*b(j);
    end
end
```

Watch the elimination process by displaying the matrix and the right hand
side after each elimination step. Use the pause statement to stop execution.

Solution:

```
for j=1:n-1                          % Elimination
    for k=j+1:n
      fak=A(k,j)/A(j,j);
      A(k,j:n)=A(k,j:n)-fak*A(j,j:n);
      b(k)=b(k)-fak*b(j);
    end
    [A, b]                           % display A and b
    pause                            % wait for ret
end
```

We observe the elimination process and get at the end

```
>> A
A =
          25        -300        1050        -1400         630
           0        1200       -6300        10080       -5040
           0           0        2205        -5880        3780
           0           0           0          448        -504
           0           0           0            0           9
>> b
b =
      1.0000
     12.0000
     21.0000
     11.2000
      1.8000
```

We see the reduction of $A$ to an upper triangular matrix $U$.

(b) Next store the factors `fak` instead of the zeros you introduce by eliminating $x_j$:

```
tor j=1:n-1                              % Elimination
  for k=j+1:n
    fak=A(k,j)/A(j,j);
    A(k,j)=fak;                          % store factors instead zeros
    A(k,j+1:n)=A(k,j+1:n)-fak*A(j,j+1:n);
  end
end
```

Now use the commands `triu, tril, diag` to extract $L$ and $U$ from $A$ and verify that indeed $LU = A$.

Solution: The second version of Gaussian Elimination stores the factors used for elimination where we would produce the zeros. Notice the difference

```
A(k,j:n)=A(k,j:n)-fak*A(j,j:n);      % first version
A(k,j+1:n)=A(k,j+1:n)-fak*A(j,j+1:n);  % second version
```

Now we get the results

```
A =
   2.5000e+01  -3.0000e+02   1.0500e+03  -1.4000e+03   6.3000e+02
  -1.2000e+01   1.2000e+03  -6.3000e+03   1.0080e+04  -5.0400e+03
   4.2000e+01  -5.2500e+00   2.2050e+03  -5.8800e+03   3.7800e+03
  -5.6000e+01   8.4000e+00  -2.6667e+00   4.4800e+02  -5.0400e+02
   2.5200e+01  -4.2000e+00   1.7143e+00  -1.1250e+00   9.0000e+00
```

The reduced matrix $U$ is still in the upper part and can be extracted by

```
>> U=triu(A)
U =
          25        -300        1050       -1400         630
           0        1200       -6300       10080       -5040
           0           0        2205       -5880        3780
           0           0           0         448        -504
           0           0           0           0           9
```

The function `triu` is an abbreviation for "upper triangle". The matrix $L$ is constructed from the factors we stored instead of the zeros. With

```
>> L=tril(A)
L =
   1.0e+03 *
    0.0250        0           0           0           0
   -0.0120    1.2000         0           0           0
    0.0420   -0.0053     2.2050         0           0
   -0.0560    0.0084    -0.0027     0.4480         0
    0.0252   -0.0042     0.0017    -0.0011      0.0090
```

we get the lower triangle of $A$ including the diagonal. We need now to replace the diagonal by all number 1. The function `diag` is useful for that. With

```
>> D=diag(L)
D =
            25
          1200
          2205
           448
             9
```

we extract the diagonal as a vector. If we use `diag` with a vector as argument
then a diagonal matrix is produced:

```
>> D=diag(D)
D =
        25           0           0           0           0
         0        1200           0           0           0
         0           0        2205           0           0
         0           0           0         448           0
         0           0           0           0           9
```

Now we can form

```
>> L=L-D+eye(5)
L =
     1.0000          0          0          0          0
   -12.0000     1.0000          0          0          0
    42.0000    -5.2500     1.0000          0          0
   -56.0000     8.4000    -2.6667     1.0000          0
    25.2000    -4.2000     1.7143    -1.1250     1.0000
```

So to summarize we just have to write

```
>>    U=triu(A);                        % decompose A, upper part
>>    L=tril(A);                        % extract lower triangular matrix
>>    L=L-diag(diag(L))+eye(size(L));   % adjust diagonal
>>    L*U
ans =
        25        -300        1050       -1400         630
      -300        4800      -18900       26880      -12600
      1050      -18900       79380     -117600       56700
     -1400       26880     -117600      179200      -88200
       630      -12600       56700      -88200       44100
```

and we get `L*U=A` as expected.

2. Replace the computation of the rotation matrix S in our function
`EliminationGivens` by the MATLAB-function `planerot`. Convince your-
self that you get the same results with the modified function by solving the curve
fitting example again.

Solution:

```
function x=EliminationGivens2(A,b);
% ELIMINATIONGIVENS solves a linear system using Givens-rotations
%    x=EliminationGivens(A,b) solves Ax=b using Givens-rotations.
```

```
[m,n]=size(A);
for i= 1:n
  for k=i+1:m
    if A(k,i)~=0
      [S,y] = planerot([A(i,i);A(k,i)]);
      A(i,i)=y(1);
      A(i:k-i:k,i+1:n)=S*A(i:k-i:k,i+1:n);
      b(i:k-i:k)=S*b(i:k-i:k);
    end
  end;
  if A(i,i)==0
    error('Matrix is rank deficient');
  end;
end
x=zeros(n,1);
for k=n:-1:1           % backsubstitution
  x(k)=(b(k)-A(k,k+1:n)*x(k+1:n))/A(k,k);
end
x=x(:);
```

Indeed by replacing `EliminationGivens` by `EliminationGivens2` in `CurveFit` we get the same results.

3. Determine the parameters $a$ and $b$ such that the function $f(x) = ae^{bx}$ fits the following data

| $x$ | 30.0 | 64.5 | 74.5 | 86.7 | 94.5 | 98.9 |
|-----|------|------|------|------|------|------|
| $y$ | 4 | 18 | 29 | 51 | 73 | 90 |

Plot the points and the fitted function.

**Hint**: If you fit log $f(x)$ the problem becomes very easy!

Solution: Taking the logarithm of the function we get

$$\ln y = \ln a + bx.$$

With the unknown $c = \ln a$ the least squares problem becomes

$$\begin{pmatrix} 1 & 30.0 \\ 1 & 64.5 \\ 1 & 74.5 \\ 1 & 86.7 \\ 1 & 94.5 \\ 1 & 98.9 \end{pmatrix} \begin{pmatrix} c \\ b \end{pmatrix} \approx \begin{pmatrix} \ln 4 \\ \ln 18 \\ \ln 29 \\ \ln 51 \\ \ln 73 \\ \ln 90 \end{pmatrix}.$$

```
% Problem 6_4_1
clear, clf
x=[30.0, 64.5, 74.5, 86.7, 94.5, 98.9]';
y=[4, 18, 29, 51, 73, 90]';
A=[ones(size(x)), x]
```

```
b=log(y);
p=A\b
a=exp(p(1))
b=p(2)
plot(x,y,'o')
hold
z=[30:100];
plot(z, a*exp(b*z))
```

The solution is

$$b = 0.04524310648 \text{ and } c = 0.00789262406 \Rightarrow a = 1.00792752.$$

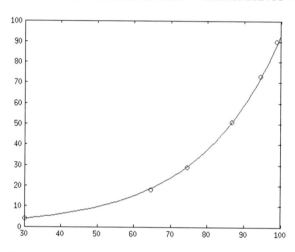

4. The following statistics lists the population of Shanghai since 1953:

| year | in million |
|------|-----------|
| 1953 | 6.2044 |
| 1964 | 10.8165 |
| 1982 | 11.8597 |
| 1990 | 13.3419 |
| 2000 | 16.4077 |
| 2010 | 23.0192 |

Fit a polynomial through these data and predict the population for 2016 and 2020. Plot your results.

Solution:

MATLAB provides the functions polyfit to fit a polynomial through points and polyval to evaluate a polynomial at some points. Thus the problem is readily solved:

```
x=[1953 1964 1982 1990 2000 2010]'
y=[6.2044 10.8165 11.8597 13.3419 16.4077 23.0192]'
p=polyfit(x,y,5)              % Interpolate by polynomial of deg. 5
xx=[x;2016; 2020]             % add extrapolation points
yy=[y;polyval(p,[2016 2020])']% extrapolate
plot(xx,yy,'*')
```

We get the result 31.5600 millions for 2014 and 40.8764 millions for 2020.

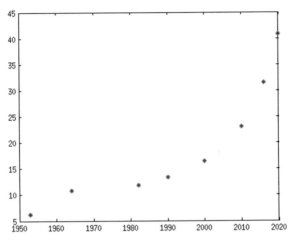

Without using the above mentioned functions we have to compute the coefficients of the interpolating polynomial by solving the system:

$$p_1 x_1^5 + p_2 x_1^4 + \cdots + p_5 x_1 + p_6 = y_1$$
$$p_1 x_2^5 + p_2 x_2^4 + \cdots + p_5 x_2 + p_6 = y_2$$
$$\vdots \qquad\qquad\qquad \vdots \ \ \vdots$$
$$p_1 x_6^5 + p_2 x_6^4 + \cdots + p_5 x_6 + p_6 = y_6$$

The matrix is a Vandermonde matrix. It can be generated by

```
A=[x.^5, x.^4, x.^3, x.^2, x, ones(size(x))]
```

So the coefficients are obtained by p=A\y. Indeed we get

```
>> A=[x.^5, x.^4, x.^3, x.^2, x, ones(size(x))]
A =
   1.0e+16 *
    2.8413   0.0015   0.0000   0.0000   0.0000   0.0000
    2.9222   0.0015   0.0000   0.0000   0.0000   0.0000
    3.0586   0.0015   0.0000   0.0000   0.0000   0.0000
    3.1208   0.0016   0.0000   0.0000   0.0000   0.0000
    3.2000   0.0016   0.0000   0.0000   0.0000   0.0000
    3.2808   0.0016   0.0000   0.0000   0.0000   0.0000
>> p=A\y
Warning: Matrix is close to singular or badly scaled. Results may be inaccurate.
RCOND =  2.726773e-28.
```

```
p =
   1.0e+09 *
    0.0000
   -0.0000
    0.0000
   -0.0000
    0.0222
   -8.7968
```

To evaluate the polynomial we write the function

```
function y=EvalPoly(p,x)
n=length(p);
y=0;
for i=1:n
  y=y*x+p(i);
end
```

and we obtain the same results as before:

```
>> y=EvalPoly(p,2016)
y =
   31.5602
>> y=EvalPoly(p,2020)
y =
   40.8768
```

**Remark**: The solution is numerically not optimal. It is not good to evaluate a polynomial in standard form for a few values far away from the origin. A numerically
better solution would be to make a shift and to work with the polynomial

$$p(x) = p_1(x - 1980)^5 + p_2(x - 1980)^4 + \cdots + p_6.$$

5. *Fitting of circles*. We are given the measured points $(\xi_i, \eta_i)$:

| $\xi$ | 0.7 | 3.3 | 5.6 | 7.5 | 6.4 | 4.4 | 0.3 | −1.1 |
|---|---|---|---|---|---|---|---|---|
| $\eta$ | 4.0 | 4.7 | 4.0 | 1.3 | −1.1 | −3.0 | −2.5 | 1.3 |

Find the center $(c_1, c_2)$ and the radius $r$ of a circle $(x - c_1)^2 + (y - c_2)^2 = r^2$ that
approximate the points as well as possible. Consider the *algebraic fit*: Rearrange
the equation of the circle as

$$2c_1 x + 2c_2 y + r^2 - c_1^2 - c_2^2 = x^2 + y^2. \tag{10.1}$$

With $w = r^2 - c_1^2 - c_2^2$, we obtain with (6.7) for each measured point a linear
equation for the unknowns $c_1$, $c_2$ and $w$.

- Write a function function drawcircle(C,r) to plot a circle with center (C(1),C(2)) and radius r.
- Computer the center and the radius and plot the given points and the fitted
  circle.

Solution:

We first write a function to plot a circle

```
function drawcircle(C,r,w);
% draws a circle with center (C(1), C(2))   and radius r
if nargin==2, w ='-'; end
theta = [0:0.02:2*pi];
plot(C(1)+r*cos(theta), C(2)+r*sin(theta),w);
plot(C(1),C(2),'x');
```

The main program is straightforward:

```
xi  = [ 0.7 3.3 5.6 7.5  6.4  4.4  0.3 -1.1]';
eta = [ 4.0 4.7 4.0 1.3 -1.1 -3.0 -2.5  1.3]';
A = [2*xi 2*eta ones(size(xi))]
b=xi.^2+eta.^2;
x = A\b
C=x(1:2)
r=sqrt(x(3)+C(1)^2+C(2)^2)
plot(xi,eta,'o')
axis equal
hold
drawcircle(C,r)
```

The results are

```
C =
    3.060303565727350
    0.743607321042322
>> r
r =
    4.109137036074778
```

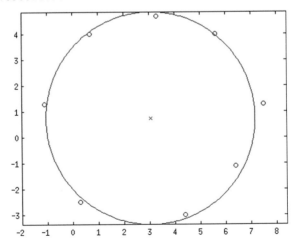

6. Seven dwarfs are sitting around a table. Each one has a cup. The cups contain milk,
   all together a total of 3 liter. One of the dwarfs starts distributing his milk evenly

over all cups. After he has finished his right neighbor does the same. Clockwise the next dwarfs proceed distributing their milk. After the 7th dwarf has distributed his milk, there is in each cup as much milk as at the beginning. How much milk was initially in each cup?

**Hint**: Let $x = (x_1, x_2, \ldots, x_7)^\top$ be the initial milk distribution. Thus $\sum_{j=1}^{7} x_j = 3$. Simulate the distributing of milk as matrix-vector Operation:

$$x^{(1)} = T_1 x.$$

After 7 distributions you obtain $x^{(7)} = x$ and thus

$$x = T_7 T_6 \cdots T_1 x$$

or $(A - I)x = 0$ where $A = T_7 T_6 \cdots T_1$. Add to this homogeneous system the equation $\sum_{j=1}^{7} x_j = 3$ and solve the system using our function Elimination Givens. Compare the results you get with those when using MATLAB's \-operator.

Solution: Let

$$x_1^{(0)}, \ldots, x_7^{(0)} \text{ be the initial milk amounts}$$

and let

$$x_1^{(1)}, \ldots, x_7^{(1)} \text{ the amounts after the first distribution.}$$

We have

$$x^{(1)} = T^{(1)} x^{(0)},$$

where

$$T^{(1)} = \begin{pmatrix} 1/7 & 0 & \cdots & & 0 \\ 1/7 & 1 & 0 & \cdots & 0 \\ 1/7 & 0 & 1 & \ddots & \vdots \\ \vdots & \vdots & \ddots & \ddots & 0 \\ 1/7 & 0 & \cdots & 0 & 1 \end{pmatrix}.$$

The $i$th distribution is given by the transformation

$$x^{(i)} = T^{(i)} x^{(i-1)}$$

where $T^{(i)}$ looks like $T^{(1)}$ only the column with the elements $1/7$ is now the $i$th column instead of the first. For the final state we have

$$x^{(7)} = \underbrace{T^{(7)} \cdots T^{(1)}}_{A} x^{(0)}$$

and since $x^{(7)} = x^{(0)} =: x$ we obtain the homogeneous linear system

$$(A - I)x = 0.$$

One can prove that a non-trivial solutions exist. The solution is unique if we consider the total amount of milk and if we add the equation

$$x_1 + x_2 + \cdots + x_7 = 3$$

to the system.

```
% Dwarfs Milk Distribution Problem
I=eye(7);
e=ones(7,1);
e7=e/7;
A=I;
for k=1:7
    T=I;                    % construct T_i
    T(:,k)=e7;
    A=T*A;
end
B=[A-I; e'];
b=[zeros(size(e));3];       % add total amount of milk
x=B\b

Check = A*x

x =
     0.7500
     0.6429
     0.5357
     0.4286
     0.3214
     0.2143
     0.1071
```

7. The following sections were measured on the street $\overline{AD}$ depicted in Fig. 6.1 (Fig. 10.2).

$$AD = 89\,m, \; AC = 67\,m, \; BD = 53\,m, \; AB = 35\,m \text{ and } CD = 20\,m$$

Balance out the measured sections using the least squares method.

$A$                  $B$                  $C$                  $D$

**Fig. 10.2** Street

Solution:

Let $x_1 = \overline{AB}$, $x_2 = \overline{BC}$ and $x_3 = \overline{CD}$. The measurements lead to the equations

$$
\begin{aligned}
x_1 + x_2 + x_3 &= 89 \\
x_1 + x_2 &= 67 \\
x_2 + x_3 &= 53 \\
x_1 &= 35 \\
x_3 &= 20
\end{aligned}
\quad \Longleftrightarrow \quad Ax = b, \quad
A = \begin{pmatrix} 1 & 1 & 1 \\ 1 & 1 & 0 \\ 0 & 1 & 1 \\ 1 & 0 & 0 \\ 0 & 0 & 1 \end{pmatrix}, \quad
b = \begin{pmatrix} 89 \\ 67 \\ 53 \\ 35 \\ 20 \end{pmatrix}
$$

Solving the system with the \-operator we get

```
>>    A = [1 1 1; 1 1 0; 0 1 1; 1 0 0; 0 0 1];
>>    b = [89; 67; 53; 35; 20];
>>    x = A\b
x =
   35.1250
   32.5000
   20.6250
```

We obtain the same result with our function EliminationGivens:

```
>> x=EliminationGivens(A,b)
x =
   35.1250
   32.5000
   20.6250
```

## 10.7   Chapter 7: Recursion

1. Cramer's Rule for solving systems of linear equations. This rule is often used when solving small ($n \leq 3$) systems of linear equations by hand.
   Write a function x=Cramer(A,b) which solves a linear system $Ax = b$ using Cramer's rule. For $\det(A) \neq 0$, the linear system has the unique solution

$$
x_i = \frac{\det(A_i)}{\det(A)}, \quad i = 1, 2, \ldots, n, \tag{10.2}
$$

   where $A_i$ is the matrix obtained from $A$ by replacing column $a_{:i}$ by $b$. Use the function DetLaplace to compute the determinants.
   Test your program by generating a linear system with known solution.

Solution:

```
function x=Cramer(A,b);
% CRAMER solves a linear System with Cramer's rule
%    x=Cramer(A,b); Solves the linear system Ax=b using Cramer's
```

```
%     rule. The determinants are computed using the function DetLaplace.

n=length(b);
detA=DetLaplace(A);
for i=1:n
  AI=[A(:,1:i-1), b, A(:,i+1:n)];
  x(i)=DetLaplace(AI)/detA;
end
x = x(:);

>> A=rand(7,7);
>> x=[1:7]';
>> b=A*x;
>> xx=Cramer(A,b);
>> norm(x-xx)
ans =
    1.1997e-13
```

2. Selection Sort versus Quick Sort.
   The idea of selection sort is to find the minimum value in the given array and then
   swaps it with the value in the first position. By repeating this for the remaining
   elements the array is sorted.

   (a) Write a (non-recursive) function a=SelectSort(a) which implements
       the Selection Sort. Show the process using bar and pause as done in Quick
       Sort. Test your program by sorting some small arrays ($n \leq 100$)

       Solution:
       ```
       function a = SelectionSort(a)
       n=length(a);
       for i=1:n-1
         [amin,k]=min(a(i:n)); k=k+i-1;
         if k~=i  % swap
             h=a(k); a(k)=a(i); a(i)=h;
       %        bar(a); pause(0.01)
         end
       end
       ```

   (b) Speed Test: Remove the bar and pause statement in both functions and
       measure the time each function needs to sort an array of 100'000 elements.
       Use for this the MATLAB-functions tic and toc.

       Solution: We comment out the line in both functions with the statements
       bar(a); pause(0.01) and run the main program

       ```
       % speed test
       clear
       global a
       n=100000
       aa=rand(1,n);
       a=aa; tic, a=SelectionSort(a); toc
       a=aa; tic, quick(1,n); toc
       ```

We get on our laptop the result

```
>> SpeedTest
n =
       100000
Elapsed time is 13.458101 seconds.
Elapsed time is 1.511749 seconds.
```

Clearly `quick` wins over `SelectionSort`.

(c) For fun (not efficient!): program the selection sort recursively. Use a global array and proceed similarly as with quicksort.

Solution: We need a main program which defines the global array $a$:

```
% SelectionSortMain.m
global a
n=50
a=rand(1,n);
bar(a)
pause
SelectionSortRec(1,n)
bar(a)
```

The recursive function `SelectionSortRec` is designed similarly like `quick`. However, it is very inefficient since it is a single recursion. For $n$ elements we need also $n$ recursive calls!

```
function  SelectionSortRec(left,n)
global a
if left<n,
   [amin,k]=min(a(left:n));
   k=k+left-1;
   if k~=left                   % swap
      h=a(k); a(k)=a(left); a(left)=h;
   end
   bar(a); pause(0.01)
   left=left+1;
   SelectionSortRec(left,n);
end
```

3. Pythagoras Tree[1]:

Basic construction: Given two points $P$ and $Q$ in the plane, construct the points $P'$ and $Q'$ to built a square. Then put on the square a right triangle with one basis angle $\alpha$.

The following figure shows the basic construction and the first recursion step, where the construction is repeated on top of the cathetes of the triangle $\overline{P'RQ'}$.

---

[1] https://en.wikipedia.org/wiki/Pythagoras_tree_(fractal).

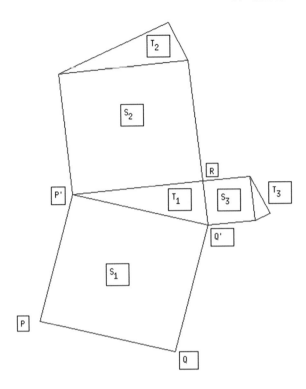

Write a recursive function which computes the Pythagoras tree until the base line
$\overline{PQ}$ becomes small. Experiment with the basis angle, choose e.g. as here in the
figure $\alpha = 20°$.

Solution: Recursive function:

```
function ptree(P,Q,alpha)
% PTREE constructs a Pythagorean tree
% ptree(P,Q,alpha) plots a Pythagorean tree over the basis line given by the
% two points P=(p1,p2) an Q=(q1,q2) in the plane with a right triangle
% with one basis angle alpha (in degrees)

g=norm(P-Q);                    % length of square
r=Q-P; r=r/norm(r);             % direction vector
n=[-r(2), r(1)];                % normal vector
Ps=P+n*g; Qs=Q+n*g;             % other corners of square
X=[P;Q;Qs;Ps;P]';
plot(X(1,:), X(2,:))            % draw the square

c=cos(alpha); s=sin(alpha);
Rot=[c s;-s c];                 % rotation matrix Rot
R=Ps+r*Rot*g*c;                 % construct triangle Ps-R-Qs over Ps-Qs
Y=[Ps;R;Qs]';                   % draw triangle
plot(Y(1,:), Y(2,:))
  if g>0.1                      % recursion for both sides
    ptree(Ps,R,alpha);          % of triangle
```

```
   ptree(R,Qs,alpha)
end
```

## Main program:

```
% Mainprogram Mainptree.m
% recursive Pythagoras tree
% example of a fractal
clf
axis([-25,15,-5,35])
axis square
hold
alpha=input('angle=? (degrees)')
ptree([1,1],[5,0],alpha*pi/180)
```

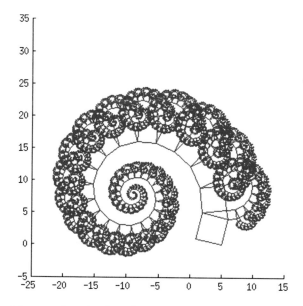

4. Permutations: MATLAB has the function `perms` to compute permutations. For instance

```
>> a=[1 2 3]
a =
       1       2       3
>> perms(a)
ans =
       3       2       1
       3       1       2
       2       3       1
       2       1       3
       1       2       3
       1       3       2
```

displays all 6 permutations of the three numbers.

Write a recursive function `Permute(n)` which does the same. Choose the array a as global variable.

Solution:

```
function Permute(k)
global a Z
if k==1                          % finish recursion
   Z=[Z;a];                      % and store permutation
else
   Permute(k-1);                 % permute a_1, ..., a_{k-1}
   for i=1:k-1
       t=a(i); a(i)=a(k); a(k)=t;  % exchange a_i <-> a_k
       Permute(k-1)                % and permute
       t=a(i); a(i)=a(k); a(k)=t;  % exchange back
   end
end
```

Indeed we get with

```
% MainPermute.m
global a Z
a=[1 2 3]
Z=[];
Permute(3)
Z

>> MainPermute
a =
     1    2    3
Z =
     1    2    3
     2    1    3
     3    2    1
     2    3    1
     1    3    2
     3    1    2
```

the desired permutations.

## 10.8   Chapter 8: Iteration and Nonlinear Equations

1. Bisection-Algorithm. Improve the function `Bisekt`. Your function `[x,y]= Bisection(f,a,b,tol)` should also compute a zero for functions with $f(a) > 0$ and $f(b) < 0$ to a given tolerance tol. Be careful to stop the iteration in case the user asks for a too small tolerance! If by the bisection process we arrive at an interval $(a, b)$ which does not contain a machine number anymore then it is high time to stop the iteration.

Solution:

```
function [x,y]=Bisection(f,a,b,tol)
% BISECTION computes a root of a scalar equation
%   [x,y]=Bisection(f,a,b,tol) finds a root x of the scalar function
%   f in the interval [a,b] up to a tolerance tol. y is the
%   function value at the solution

fa=f(a); v=1; if fa>0, v=-1; end;
if fa*f(b)>0
  error('f(a) and f(b) have the same sign')
end
if (nargin<4), tol=0; end;
x=(a+b)/2;
while (b-a>tol) & ((a < x) & (x<b))
  if v*f(x)>0, b=x; else a=x; end;
  x=(a+b)/2;
end
if nargout==2, y=f(x); end;
```

2. Solve with bisection the equations

$$(a) \quad x^x = 50 \qquad (b) \quad \ln(x) = \cos(x) \qquad (c) \quad x + e^x = 0.$$

**Hint**: a starting interval is easy to find by sketching the functions involved.

Solution:

(a) The function $x^x$ is monotonically increasing. Since $1^1 = 1$ and $4^4 = 256$ the values $a = 1$ and $b = 4$ can be used for the bisection. The solution becomes

```
>> [x,f]=Bisection(@(x) x^x-50,1,4)
x =
    3.287262195355581
f =
      7.105427357601002e-15
```

(b) Drawing the functions $\ln(x)$ and $\cos(x)$ we see that their cutting point is in the interval $(0, \pi/2)$, thus

```
>> [x,f]=Bisection(@(x) log(x)-cos(x),0,pi)
x =
   1.302964001216012
f =
     -2.220446049250313e-16
```

(c) We write the equation $e^x = -x$ and from the graph of the two functions we get the interval $(-1, 0)$ for the solution, so

```
>> [x,f]=Bisection(@(x) exp(x)+x,-1,0)
x =
  -0.567143290409784
f =
     -1.110223024625157e-16
```

3. Find $x$ such that

$$\int_0^x e^{-t^2} dt = 0.5.$$

**Hint**: the integral cannot be evaluated analytically, so expand it in a series and integrate. Write a function $f(x)$ to evaluate the series. Then use bisection to compute the solution of $f(x) - 0.5 = 0$.

Solution:

Take the series for $e^x$, substitute $x = -t^2$ and integrate to get the expansion

$$\int_0^x e^{-t^2} dt = x - \frac{x^3}{1! \, 3} + \frac{x^5}{2! \, 5} - \frac{x^7}{3! \, 7} + \frac{x^9}{4! \, 9} \mp \cdots \qquad (10.3)$$

For evaluating the series we introduce the expressions

$$ta := (-1)^{i-1} \frac{x^{2i-1}}{(i-1)!} \qquad t := (-1)^i \frac{x^{2i+1}}{i!}$$

then $t = -ta * x^2/i$ and the partial sum is updated by $s_{new} = s_{old} + t/(2*i+1)$. We will stop the summation when $s_{new} = s_{old}$. Thus we get

```
function y=ff(x);
% is used in IntegralExp.m
t=x; snew=x; sold=0; i=0;
while sold ~= snew
  i=i+1;
  sold=snew;
  t=-t*x^2/i;
  snew=sold+t/(2*i+1);
end
y=snew;

>> [x,f]=Bisection(@(x) ff(x)-0.5,0,1)
x =
    0.551039427609027
f =
    0
```

4. Use bisection to create the following table:

| F | 0 | $0.1\pi$ | $0.2\pi$ | ... | $\pi$ |
|---|---|----------|----------|-----|-------|
| h | 0 | ? | ? | ... | 2 |

where the function $F(h)$ is given by

$$F(h) = \pi - 2 \arccos \frac{h}{2} + h\sqrt{1 - \left(\frac{h}{2}\right)^2}.$$

Solution: To compute the table entries we loop through the values $0 : 0.1\pi : \pi$. The equation to be solved by bisection changes during the loop. A simple way to deal with this is to use a global variable. The main program becomes:

```
% Table.m
global f
format long
res=[];
for f=0:0.1*pi:pi;
  [x y]=Bisection(@F,0,2);
  res=[res; f x y];
end;
res
```

and the function is

```
function  z=F(h)
global f
z=pi-2*acos(h/2.0)+h.*sqrt(1-(h/2).^2)-f;
```

Running the program we obtain

```
res =
                 0                   0                   0
 0.314159265358979   0.157241774836456   0.000000000000000
 0.628318530717959   0.315472387600032                   0
 0.942477796076938   0.475764878639765  -0.000000000000000
 1.256637061435917   0.639383019581008   0.000000000000000
 1.570796326794897   0.807945506599034  -0.000000000000000
 1.884955592153876   0.983723665527420   0.000000000000000
 2.199114857512855   1.170274845761590                   0
 2.513274122871834   1.374097652265081   0.000000000000000
 2.827433388230814   1.610767273040240                   0
 3.141592653589793   2.000000000000000                   0
```

5. *Binary search*: we are given an ordered sequence of numbers:

$$x_1 \leq x_2 \leq \cdots \leq x_n$$

and a new number $z$. Write a program that computes an index value $i$ such that either $x_{i-1} < z \leq x_i$ or $i = 1$ or $i = n + 1$ holds. The problem can be solved by considering the function

$$f(i) = x_i - z$$

and computing its "zero" by bisection.

Solution:

```
function Binsearch(z,x)
n=length(x);
if z<x(1),
    disp(strcat('z = ',num2str(z),' is smaller than   x(1) = ',num2str(x(1))))
elseif z>x(n),
    disp(strcat('z = ',num2str(z),' is larger than   x(n) = ',num2str(x(n))))
else
    a=1; b=n;
    while a+1~=b
      i=round((a+b)/2);
      if x(i)<z; a=i;
      else b=i;
      end
    end
    i=b;
    disp(strcat('i = ',num2str(i),' and ','    z = ',num2str(z),' is in [',...
      num2str(x(a)),', ',num2str(x(b)),']'))
end
```

We test this function with the main program

```
% MainBinSearch.m
x=[-1.4 0 3.7 5.1 7.9 9.4 11.6 13.1 17 25.4 26]'
for z=[8, 17.5, -3, 32]
  BinSearch(z,x)
end

>> MainBinSearch
x =
    -1.4000
         0
     3.7000
     5.1000
     7.9000
     9.4000
    11.6000
    13.1000
    17.0000
    25.4000
    26.0000
i =6 and   z =8 is in [7.9,9.4]
i =10 and   z =17.5 is in [17,25.4]
z =-3 is smaller than   x(1) =-1.4
z =32 is larger than   x(n) =26
```

6. Compute $x$ where the following maximum is attained:

$$\max_{0<x<\frac{\pi}{2}} \left( \frac{1}{4\sin x} + \frac{\sin x}{2x} - \frac{\cos x}{4x} \right).$$

Solution: We differentiate and compute the zero of the derivative using bisection. We choose for $a = 0.001$ since with $a = 0$ the denominator is zero

```
>> Bisection(@(x) cos(x)/4/x*(2-x/(sin(x))^2+1/x)+...
      sin(x)/4/x*(1-2/x),0.001,pi/2)
ans =
    1.031158096685125
```

7. Write a function `s=SquareRoot(a)` which computes the square root using
Heron's algorithm. Think of a good starting value and a good termination criterion.

**Hint**: consider the geometrical interpretation of Newton's method and use the
(theoretical) monotonicity of the sequence as termination criterion.
Test your function and compare the results with the standard MATLAB-function
`sqrt`. Compute the relative error of both functions

Solution:

```
function s = SquareRoot(a);
% SQUAREROOT computes the square root
%   using Heron's algorithm
xold = (1+a)/2; xnew = (xold+a/xold)/2;
while xnew<xold       % as long as monotone
  xold = xnew; xnew = (xold+a/xold)/2;
end
s = xnew;
```

With the testprogram

```
% TestSquareRoot.m
z=[];
for x = 1:10:1000
  z = [z;  x  (SquareRoot(x)-sqrt(x))/sqrt(x) ];
end
for i = 1:30
  fprintf('%10d %15.6e %10d %15.6e %10d %15.6e\n', z(i,1),z(i,2), ...
          z(i+30,1),z(i+30,2), z(i+60,1),z(i+60,2) )
end
```

we get excellent results.

8. We consider again Problem 3: find $x$ such that

$$f(x) = \int_0^x e^{-t^2} dt - 0.5 = 0.$$

Since a function evaluation is expensive (summation of the Taylor series) but the
derivatives are cheap to compute, a higher order method is appropriate. Solve this
equation with Newton's method.

Solution:

```
% IntegralExp.m
% Solve \int_{0}^{x} e^{-t^2}dt - 0.5 = 0 with Newton
```

```
% use ff.m to compute Taylor series
format compact, format long
x=1; xa=2;
while abs(xa-x)>1e-10
  xa=x;
  y=ff(x)-0.5; ys=exp(-x^2);
  x=x-y/ys
end

>> IntegralExp
x =
   0.329062444950818
x =
   0.532365165339031
x =
   0.550852862865461
x =
   0.551039408434969
x =
   0.551039427609027
x =
   0.551039427609027
```

9. Using Newton's iteration, find $a$ such that $\displaystyle\int_0^1 e^{at}\,dt = 2$.

    Solution:

    We have

    $$\int_0^1 e^{at}\,dt = \frac{1}{a}e^{at}\Big]_0^1 = \frac{1}{a}e^a - \frac{1}{a}$$

    Thus we are looking for the solution of the equation

    $$f(a) = \frac{e^a - 1}{a} - 2 = 0$$

    We use Newton's method and get $a = 1.2564312086$.

10. Consider the billiard-problem. Let the ball $P$ be at position $P = (0.5, 0.5)$ and let $Q$ move in small steps (say 0.1) from 1 to $-1$.
    Compute for each position the solutions using bisection. Count and plot the solutions and plot also the function billiard. make a pause before moving on the the next position of $Q$.

    Solution:

```
% MainBilliard.m
% Animation for the billiard problem
clear, clf, format compact
figure(1)                     % to show the billiard table
```

```
  figure(2)                    % to plot the function billiard(x)
  pause                        % to separate the two figures
  global px py a               % ball positions P=(px.py), Q=(a,0)

  px=0.5;  py=0.5;

  for a=1:-0.1:-1              % move Q
    a
    figure(1), clf(1)
    axis equal, hold
    t=0:0.01:2*pi;             % plot circle
    plot(cos(t),sin(t))
    plot(px,py,'o')            % plot point P
    text(px,py,'    P')
    plot(a,0,'o')              % plot point Q
    text(a,0,'   Q')
    P=[px,py]; Q=[a,0];
  %%%%%%%%%%%%%%%%%%%%%%%%%%%%%%%%%%%%%%%%%%%%%%%%%%%%%%%%%%%%%%%%%
    N=200; h=2*pi/N;                       % sample billiard
    xa=0; fa=billiard(xa);
    k=0; Sols=[];
    for i=1:N,
      xb=i*h; fb=billiard(xb);
      if fa*fb <=0;                        % if sign change call Bisection
        x=Bisection(@billiard,xa,xb)       % compute angle
        k=k+1; Sols=[Sols,x];              % count and store solution
        X=[cos(x),sin(x)]                  % Reflection point
        text(X(1),X(2),['  X',num2str(k)]);
        plot([Q(1),X(1)], [Q(2),X(2)])     % plot trajectory
        plot([X(1),P(1)], [X(2),P(2)])
      end;
      xa=xb; fa=fb;
    end;
  %%%%%%%%%%%%%%%%%%%%%%%%%%%%%%%%%%%%%%%%%%%%%%%%%%%%%%%%%%%%%%%%%
    F=[];
    X=0:0.01:2*pi;
    for x=X
      F=[F,billiard(x)];
    end
    figure(2)
    plot(X,F,[0,2*pi],[0,0])
    legend('billiard(x)')
    Sols
    k
    if a==1, pause      % to explain what is going on
    else
      pause(1)
    end
  end
```

11. Modify the fractal program by replacing $f(z) = z^3 - 1$ with the function

$$f(z) = z^5 - 1.$$

(a) Compute the 5 zeros of $f$ using the command `roots`.
(b) In order two distinguish the 5 different numbers, study the imaginary parts of the 5 zeros. Invent a transformation such that the zeros are replaced by 5 different positive integer numbers.

Solution: We first compute the zeros of $z^5 - 1$. The coefficients of the polynomial $z^5 - 1$ are

```
p=[1 0 0 0 0 -1]
```

With the function `roots` we can compute the zeros

```
>> W=roots(p)
W =
  -0.809016994374948 + 0.587785252292473i
  -0.809016994374948 - 0.587785252292473i
   0.309016994374947 + 0.951056516295152i
   0.309016994374947 - 0.951056516295152i
   1.000000000000000 + 0.000000000000000i
```

If we multiply the imaginary part by 2 we get

```
>> 2*imag(W)
ans =
   1.175570504584946
  -1.175570504584946
   1.902113032590305
  -1.902113032590305
                   0
```

Now we can add 3 and round the result to get

```
>> round(2*imag(W)+3)
ans =
     4
     2
     5
     1
     3
```

By multiplying with the factor $a = 10$ we get a beautiful fractal.

```
n=1000; m=30;
x=-1:2/n:1;
[X,Y]=meshgrid(x,x);
Z=X+1i*Y;                          % define grid for picture
for i=1:m                          % perform m iterations in parallel
  Z=Z-(Z.^5-1)./(5*Z.^4);          % for all million points
end;                               % each element of Z contains one root
```

```
a=10;
image(round(2*imag(Z)+3)*a);
```

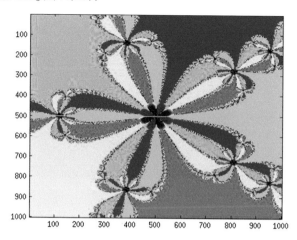

12. Mandelbrot set: Consider the iteration

$$Z_{k+1} = Z_k^2 + C.$$

Depending on the value of the constant $C$ the sequence $\{Z_k\}$ will either diverge to $\pm\infty$ or converge.
Let $C$ now be in the region in the complex plane $Z = X + iY$ with $-2 \leq X, Y \leq 2$.
Perform 50 iterations starting always with $Z_0 = 0$ with all numbers $C$ in that region and plot using `image` the resulting Mandelbrot set, which is the set of all values $C$ for which the iterations converges to a finite limit.

Solution:

```
% Mandelbrot.m
clf;
n=1000; m=50;
x=-1.6:2/n:1.6;
[X,Y]=meshgrid(x,x);
C=X+1i*Y;                        % define grid for picture
Z=0*C;
for i=1:m                        % perform m iterations in parallel
  Z=Z.^2+C;                      % for all million points
end;                             % each element of Z contains the limit
a=30;
image(isfinite(Z)*a);            % and display image
```

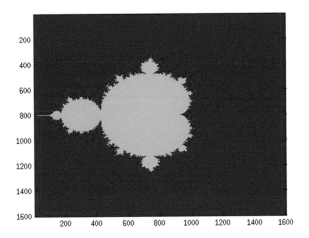

## 10.9  Chapter 9: Simulation

1. Waiting for the elevator. We consider a building with $n$ floors. A elevator is serving these floors and we are interested to know the distribution of the waiting time from pressing the elevator button till the elevator opens the door.

   Assume the time unit for the moving of the elevator one floor is one. We make $m$ experiments in which the elevator is randomly located on one floor and the person is also coming randomly on one floor. The difference of the two floors is proportional to the time the elevator needs to come.

   Perform $m = 10,000$ experiments for a $n = 10$ floors building. Construct and plot the histogram of the waiting times.

   What is the most likely waiting time?

   Solution:

```
% Waiting for the elevator
clear, clf,clc
m=10000              % number of persons
n=10                 % number of floors
A=zeros(1,n+1);
for j=1:m
  LocationElevator=round(rand*n);
  LocationPerson=round(rand*n);
  waiting=abs(LocationElevator-LocationPerson);
  A(waiting+1)=A(waiting+1)+1;
end
bar(A/m)
```

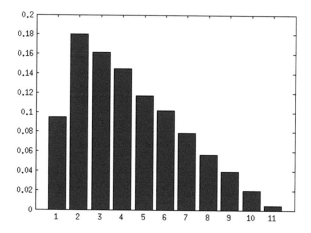

It seems that most likely the elevator is one floor up or down when a person wants to use it.

2. Given a set of points in the unit square. Write a program which computes and plots the two closest points.

   (a) Write a function `[P,Q,minimum]=ClosestPoints(x,y)` which computes all the distances between two points and stores the minimal distance and the two points $P$ and $Q$ which are closest.

   (b) Generate $n$ points $(x_k, y_k)$ using the function `rand`. Then call the function `ClosestPoints`, plot the points and mark the two closest points by coloring them differently.

Solution:

```
function [P,Q,minimum]=ClosestPoints(x,y)
% CLOSESTPOINTS computes the two closest point
%
n=length(x);
minimum=Inf;
for k=1:n
  for p=k+1:n
    dist=norm([x(k)-x(p);y(k)-y(p)]);
    if dist<minimum
      minimum=dist;
      kmin=k; pmin=p;
    end
  end
end
P=[x(kmin);y(kmin)];
Q=[x(pmin);y(pmin)];
```

for $n = 20$ we get for instance

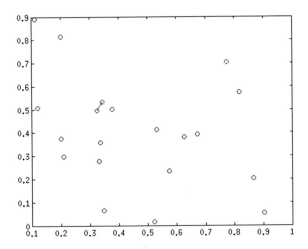

3. Shortest distance between two point sets:

   (a) Consider the circle with center $(5, 6)$ and radius $r = 2$ and the ellipse with
   center at origin and $a = 1$ and $b = 0.5$ parallel to the coordinate axis.
   Sample points on the ellipse and on the circle. Compute by brute force a
   point $P$ on the circle and a point $Q$ on the ellipse with minimal distance.
   Solution:

```
% DistEllipseCircle.m
% distance between a circle and an ellipse
clear,clf
t=linspace(0,2*pi,300);
X=[5+2*cos(t); 6+2*sin(t)];      % define circle
Y=[cos(t);0.5*sin(t)];           % define ellipse
axis([-2,10,-2,10])
axis square
hold
plot(X(1,:),X(2,:))
plot(Y(1,:),Y(2,:))
minimum=Inf;                     % compute points with
for x=X                          % minimal distance
   for y=Y
      dist=norm(x-y);
      if dist<minimum
         minimum=dist;
         P=x;Q=y;
      end
   end
end
plot([P(1),Q(1)], [P(2),Q(2)],'or')
plot([P(1),Q(1)], [P(2),Q(2)],'r')
distance=minimum
P,Q

>> DistEllipseCircle
```

```
Current plot held
distance =
     5.0796
P =
     3.8126
     4.3907
Q =
     0.8200
     0.2862
```

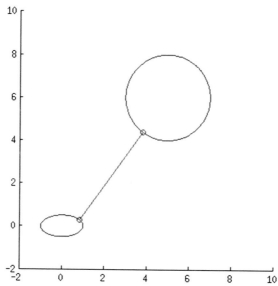

(b) The circle with center $(5, 6)$ and radius $r = 2$ and the ellipse with center $(4, 4)$, $a = 2$ and $b = 3$ intersect. Try to compute the intersection points by brute force.

Solution:

```
% IntersectEllipseCircle.m
% intersection points between a circle and an ellipse
clear,clf
t=linspace(0,2*pi,300);
X=[5+2*cos(t); 6+2*sin(t)];    % define circle
Y=[4+2*cos(t);4*3*sin(t)];     % define ellipse

axis([-2,10,-2,10])
axis square
hold
plot(X(1,:),X(2,:))
plot(Y(1,:),Y(2,:))

minimum=Inf;                   % compute points with
for x=X                        % minimal distance
   for y=Y
     dist=norm(x-y);
```

```
     if dist<0.07                        % 0.07 is by experiment
       dist
       P=x;Q=y;  [x,y]
       plot([P(1),Q(1)],  [P(2),Q(2)],'or')
     end
   end
end


>> IntersectEllipseCircle
Current plot held
dist =
    0.0686
ans =
    5.5295       5.5109
    7.9286       7.8627
dist =
    0.0304
ans =
    5.8447       5.8737
    4.1871       4.1963
dist =
    0.0127
ans =
    5.8826       5.8737
    4.2053       4.1963
dist =
    0.0542
ans =
    5.9201       5.8737
    4.2242       4.1963
```

We obtain 3 times the approximations for the second intersection point.

4. Knapsack Problem: given a bag with a given maximum load limit $W$. Put in that bag items from the following table in order to maximize the sum of the value of the items but not exceeding the total weight $W$:

| item | 1 | 2 | 3 | 4 | 5 | 6 | 7 |
|---|---|---|---|---|---|---|---|
| weights | 3.3 | 4.6 | 1.7 | 5.8 | 7.7 | 3.1 | 5.3 |
| values | 7 | 9 | 5 | 12 | 14 | 6 | 12 |

Write a brute force program that solves the problem for a collections of bags:

$$W = [8, 10, 11, 15, 20, 21, 25, 26, 30, 32]$$

Solution:

```
% knapsack problem solved by brute force
clear, clf,clc
A=[1 2 3 4 5 6 7             % item number
   3.3 4.6 1.7 5.8 7.7 3.1 5.3   % weights
     7 9 5 12 14 6 12];      % values
BagSizes=[8,10,11, 15,20,21,25,26,30,32];

Values=[]; Weights=[];
P=perms(1:length(A));
[m,n]=size(P);
for W=BagSizes
  maximum=0;
  for k=1:m
    t=0; jj=n;                % inspect row k
    for j=1:n                 % compute weights
      s=t;                    % not exceeding W
      t=s+A(2,P(k,j));
      if t>W
        jj=j-1; break
      end
    end
    s=0;
    for j=1:jj
      s=s+A(3,P(k,j));        % compute value
    end                       % and compare if larger
    if s>maximum              % than current maximum
      maximum=s; row=k; col=jj;
    end
  end
  Sol=A(:,P(row,1:col))
  weight=sum(A(2,P(row,1:col)))
  value=sum(A(3,P(row,1:col)))
  Weights=[Weights, weight];
  Values=[Values, value];
  W
```

```
    pause
end

plot(BagSizes,[Weights;Values],'o')
hold
plot(BagSizes,BagSizes)
```

We get the figure below where we see that the larger bags are the more also the total value increases. Furthermore the total weight remains always below the capacity of the bag indicated by the solid line.

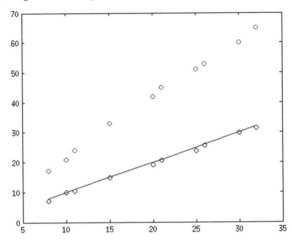

5. A dog would like to cross a river of width $b$. He starts at point $(b, 0)$ with the goal to swim to $(0, 0)$ where he has detected a sausage. His swim velocity $v_D$ is constant and his nose points always to the sausage. The river flows north in direction of the $y$-axis and velocity of the flow of the river $v_R$ is everywhere constant.

(a) Develop the differential equation describing the orbit $z(t) = (x(t), y(t))^T$ of the dog.

(b) Program a MATLAB function zp=dog(t,z) which describes the differential equation. The velocities $v_D$ and $v_R$ may be declared as global variables.

(c) Use the program quiver and plot the slope field for $b = 1$, $v_R = 1$ and the following three cases for the dog velocity $v_D = 0.8$, 1.0 and 1.5.
Note: quiver(X,Y,Xp,Yp) needs 4 matrices. X and Y contain the coordinates of the points and Xp and Yp the two components of the velocity at that point. To compute these you can use the function dog e.g.

```
z=dog(0,[X(k,j),Y(k,j)]); Xp(k,j)=z(1); Yp(k,j)=z(2);
```

(d) Develop a MATLAB integrator for the method of Heun of order 2

```
function Z= OdeHeun(f,z0,tend,n)
% ODEHEUN integrates y'=f(t,y), y(0)=z0 with Heun from
%      t=0 to tend using a fixed step size h=tend/n
```

which integrates a given system of differential equations $y' = f(t, y)$ and stores the results in the matrix $Z$. The $i$th row of the matrix $Z$ contains the values

$$[t_i, y_1(t_i), \ldots, y_n(t_i)].$$

Compute and plot the orbits for the three dog velocities. You may want to stop the integration before executing all $n$ steps when the dog arrives close to the origin or in the case when $v_D < v_R$ the dog is near the $y$-axis.

Solution:

(a) Let $z(t) = (x(t), y(t))^\top$ denote the position of the dog. The direction of the velocity vector of the dog points always to the the sausage at the origin. The velocity is overloaded with the river flow velocity which points north. Thus we get the system of differential equations:

$$z'(t) = \begin{pmatrix} x'(t) \\ y'(t) \end{pmatrix} = -\frac{v_D}{\sqrt{x(t)^2 + y(t)^2}} \begin{pmatrix} x(t) \\ y(t) \end{pmatrix} + \begin{pmatrix} 0 \\ v_R \end{pmatrix}$$

$$= -v_D \frac{z(t)}{\|z(t)\|} + \begin{pmatrix} 0 \\ v_R \end{pmatrix}$$

(b) The MATLAB function becomes

```
function zp=dog(t,z)
% dog, river problem
global vD vR
  zp=-vD*z/norm(z)+[0,vR];
end
```

(c) The program for the slope field is

```
% Velocity field of RiverDog C.m
clear, clf
global vD vR
vR=1; vD=1.5;
r=(0:.06:1)
[X,Y]=meshgrid(r,r)
[m,n]=size(X);
Xp=zeros(m,n); Yp=zeros(m,n);
for k=1:m
    for j=1:n
        z=dog(0,[X(k,j),Y(k,j)]);
        Xp(k,j)=z(1);
        Yp(k,j)=z(2);
    end
end
quiver(X,Y,Xp,Yp)
```

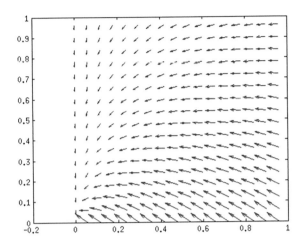

(d)  The Heun integrator becomes

```
function Z= OdeHeun(f,z0,tend,n)
% ODEHEUN integrates y'=f(t,y), y(0)=z0 with Heun from
%      t=0 to tend using a fixed step size h=tend/n
global vR vD
y=z0; t=0; Z=[t,y];
h=tend/n;
for k=1:n
  k1=f(t,y); ys=y+h*k1;
  k2=f(t+h,ys);
  y= y+h/2*(k1+k2);
  t=t+h;
  Z=[Z; t y];
  if norm(y)<0.02 | (abs(y(1))<0.02 & vR>vD)
     break
  end
end
```

we stop the integration when the dog is near the $y-axis$ or near the origin.
The main program is

```
% RiverDog.m
clear,clf
global vD vR
vR=1;
%vD=0.8
vD=1
%vD=1.5
axis([-0.1,1,-0.1,1])
axis equal
hold on
z0=[1,0];
Z=OdeHeun(@dog,z0,10,500);
plot(Z(:,2),Z(:,3))
```

and the results for some dog velocities VD are

$v_D = 0.8$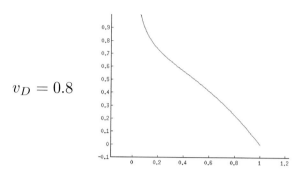

$v_D = 1$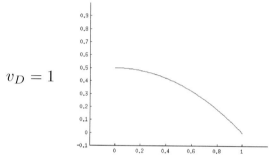

$v_D = 1.5$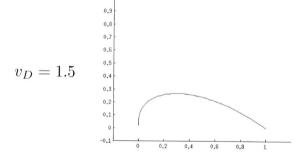

# Bibliography

1. Anderson, E., Bai, Z., Bischof, C., Demmel, J., Dongarra, J., Du Croz, J., Greenbaum, A., Hammarling, S., McKenney, A., Ostrouchov, S., Sorensen, D.: LAPACK Users Guide—Release 2.0. SIAM, Philadelphia (1992)
2. Dongarra, J.J., Bunch, J.R., Moler, C.B., Stewart, G.W.: LINPACK Users' Guide. SIAM, Philadelphia (1979)
3. Gander, W., Gander, M.J., Kwok, F.: Scientific Computing Using Maple and Matlab. Springer, Heidelberg (2014)
4. Gander, W., Hřebíček, J.: Solving Problems in Scientific Computing Using Maple and Matlab, 4th edn. Springer, Heidelberg (2004)
5. Garbow, B.S., Boyle, J.M., Dongarra, J.J., Moler, C.B.: Matrix Eigensystem Routines EISPACK Guide Extension. Lecture Notes in Computer Science. Springer, Berlin (1977)
6. Grau, A.A., Hill, U., Langmaack, H.: Translation of ALGOL 60. Springer, Berlin (1967)
7. Moler, C.B.: MATLAB Users Guide. Report. University of New Mexico, USA (1980)
8. Moler, C.B.: Experiments with matlab, on line tutorial. http://www.mathworks.com/moler/exm/chapters.html (2011)
9. Quarteroni, A., Saleri, F., Gervasio, P.: Scientific Computing with MATLAB and Octave. Springer, Berlin (2014)
10. Rutishauser, H.: Description of ALGOL 60. Springer, Berlin (1967)
11. Smith, B.T., Boyle, J.M., Dongarra, J.J., Garbow, B.S., Ikebe, Y., Klema, V.C., Moler, C.B.: Matrix Eigensystem Routines—EISPACK Guide. Lecture Notes in Computer Science. Springer, Berlin (1976)
12. Wilkinson, J., Reinsch, C.: Linear Algebra. Springer, Berlin (1971)
13. Wirth, N.: Algorithms and Data Structures. Prentice Hall, New Jersey (1986)

© Springer International Publishing Switzerland 2015
W. Gander, *Learning MATLAB*, UNITEXT - La Matematica per il 3+2 95,
DOI 10.1007/978-3-319-25327-5